The Einstein, Podolsky, and Rosen Paradox

in Atomic, Nuclear, and Particle Physics

The Einstein, Podolsky, and Rosen Paradox

in Atomic, Nuclear, and Particle Physics

Alexander Afriat
London School of Economics
London, England

and

Franco Selleri
University of Bari
Bari, Italy

Plenum Press • **New York and London**

Library of Congress Cataloging-in-Publication Data

Afriat, Alexander.
    The Einstein, Podolsky, and Rosen paradox in atomic, nuclear,
and particle physics / Alexander Afriat and Franco Selleri.
        p.   cm.
    Includes bibliographical references and index.
    ISBN 0-306-45893-4
    1. Einstein-Podolosky-Rosen experiment. 2. Nuclear physics.
I. Selleri, Franco.   II. Title.
QC174.12.A38   1998
530.12--dc21   98-43177
                                                                     CIP

ISBN 0-306-45893-4

© 1999 Plenum Press, New York
A Division of Plenum Publishing Corporation
233 Spring Street, New York, N.Y. 10013

http://www.plenum.com

Printed in the United States of America

Preface

"Paradox" conjures up arrows and tortoises. But it has a speculative, *gedanken* ring: no one would dream of really conjuring up Achilles to confirm that he catches the tortoise. The paradox of Einstein, Podolsky, and Rosen, however, is capable of empirical test. Attempted experimental resolutions have involved photons, but these are not detected often enough to settle the matter. Kaons are easier to detect and will soon be used to discriminate between quantum mechanics and local realism.

The existence of an objective physical reality, which had disappeared behind the impressive formalism of quantum mechanics, was originally intended to be the central issue of the paradox; locality, like the mathematics used, was just assumed to hold. Quantum mechanics, with its incompatible measurements, was born rather by chance in an atmosphere of great positivistic zeal, in which only the obviously measurable had scientific respectability. Speculation about occult "unobservable" quantities was viewed as vacuous metaphysics, which should surely form no part of a mature scientific attitude. Soon the "unmeasurable," once only disreputable, vanished altogether. One had first been told not to worry about it; then, as dogma got more carefully defined, one was assured that the unobserved was just not there. This made it easier not to think about it and to avoid hazardous metaphysical temptation.

If a theory indicates, in such a climate, that two quantities cannot be measured at the same time, a first, moderate reaction, is exclusive interest in one or the other. Only one can be "implemented" at a time, so why worry about both together? The next step, once one has got used to the first, is to clear up the ontological background by doing away with the complementary quantity altogether. It is wrong to say that it had a precise reality before measurement, for the situation before measurement was unmeasured. What little there is of reality has to be invoked by measurement, and complementary realities cannot be invoked together.

Common sense suggested that measurement revealed the value of one quantity in an imperfectly described reality while it disturbed others; according to dogma, on the other hand, measurement implemented one aspect of a nebulous but completely described reality to the detriment of complementary features. With a single particle the matter could not be decided, so another particle was introduced. The *reality criterion* of Einstein, Podolsky, and Rosen allowed *elements of reality* corresponding to incompatible descriptions to be assigned to an object through potential measurements on another.

With contiguous objects, the implementation of an aspect of one might cause the implementation of a corresponding aspect of the other. According to quantum mechanics, however, elements of reality and incompatible descriptions can be attributed by a measurement made at any distance. At first, the possibility of action at a distance was hardly even considered or was dismissed as necromancy: "That would be magic," as Schrödinger said. This magic was, however, soon enthusiastically embraced by many, who welcomed it as an exotic influence that entangled the constituents of a world whose workings would otherwise have been too straightforward. Of course Einstein not only believed that the world really existed, but did not accept such nonlocality:

> If one asks what, irrespective of quantum mechanics, is characteristic of the world of ideas of physics, one is first of all struck by the following: the concepts of physics relate to a real external world, that is, ideas are established relating to things such as bodies, fields, etc., which claim a "real existence" that is independent of the perceiving subject.... It is a further characteristic of these physical objects that they are thought of as arranged in a space time continuum. An essential aspect of this arrangement of things in physics is that they may claim, at a certain time, to an existence independent of one another, provided these objects 'are situated in different parts of space'. Unless one makes this kind of assumption about the independence of the existence of objects which are far apart from one another in space . . . physical thinking in the familiar sense would be impossible . . . . The following idea characterises the relative independence of objects far apart in space (*A* and *B*): external influence on *A* has no direct influence on *B*; this is known as the 'principle of contiguity' . . . . If this axiom were to be abolished . . . the postulation of laws which can be checked empirically in the accepted sense, would become impossible.[1]

If the various parts of the world influenced each other instantaneously, no regularities could be observed, no laws established.

The existence of an objective reality is becoming more and more difficult to contest. The very practice of physics, chemistry, biochemistry, neurophysiology, etc., strongly supports the reality of objects, for it is now understood how signals that give rise to perceptions are emitted, how they are propagated, and, at least partially, how neurophysiological processes transform the signals into perceptions. The realist postulate so understood is hardly arbitrary because the active relation between the subject and the world in which he operates establishes, in a sense, the existence of an objective reality *a posteriori*. Realism was characterized by de la Peña and Cetto[2] as follows:

Realism is a philosophical term to which there correspond many nonequivalent notions.... In its broad ontological meaning, (objective) realism postulates that independently of our theories and prior to them, there is an objective reality; in other words, it posits the existence of an independent reality which precedes any effort to disclose it. The task of scientific endeavour is just to disclose the nature of this reality and the laws of behaviour of its things. On the epistemological plane, realism opposes subjectivism; however, there is a rich variety of epistemologic versions of realism.

Bell derived his celebrated inequality, however, from realism and locality, and showed that it is grossly violated by quantum mechanics. So it is impossible to accept realism, locality, *and* quantum mechanics. The fact that experiments performed to decide the matter have not refuted a genuine (that is, weak) Bell inequality is well known to the experts, but there is a widespread belief that this is only due to the nonideality of the measuring devices (in particular the low efficiency of available photon counters). It is also believed that this difficulty will eventually be overcome[3]:

I always emphasize that the Aspect experiment is too far from the ideal in many ways—counter efficiency is only one of them. And I always emphasize that there is therefore a big extrapolation from practical present-day experiments to the conclusion that nonlocality holds. I myself choose to make the extrapolation, for the purpose at least of directing my own future researches. If other people choose differently, I wish them every success and I will watch for their results. . . .

The very possibility of applying the reality criterion of Einstein *et al.* and deducing Bell's inequality represents a kind of *ontological* violation, not only of complementarity but also of Heisenberg's uncertainty relations. It is a violation accepted only by realists, however, because[4]:

The uncertainty principle "protects" quantum mechanics. Heisenberg recognized that if it were possible to measure momentum and position simultaneously with a greater accuracy, quantum mechanics would collapse. So he proposed that it must be impossible. Then people sat down and tried to figure out ways of doing it, and nobody could figure out a way to measure the position of anything—a screen, an electron, a billiard ball, anything—with any greater accuracy. Quantum mechanics maintains its perilous but still accurate existence.

As noted by Croca[5] the situation is rapidly changing. The theoretical resolution limit of a microscope was established in the late XIXth century by Abbe and Rayleigh from diffraction theory as half wavelength ($\lambda/2$). The basic working of these microscopes is a fundamental example of the validity of the Heisenberg uncertainty relations. In the middle of the 80s this picture changed drastically with the development of a new generation of microscopes that in practice violate Abbe's theoretical barrier. The new generation of microscopes is typified by the scanning tunnelling electron microscope, for which Binnig and Rohrer[6] received the Nobel prize. The scanning tunnelling microscope triggered a whole variety of Scanning Probe Microscopes (SPM), where the word "Probe" can be replaced by Force (SFM), Capacitance (SCM) and Ion Conductance (SICM), opening a new era for the study of the micro world. Similar achievements were also obtained in the optical domain. In 1984 Pohl et al.[7] were able to demonstrate the feasibility of a scanning apertureless optical microscope based on Near-field Optics (SNOM) with a spatial resolution of $\lambda/20$. Ten years later it was possible to attain a resolution of $\lambda/50$ or even better.[8] These authors did not consider the implications of their accomplishments for fundamental physics, but Croca stressed that the way leading to a violation of Heisenberg's inequalities acceptable by all physicists, whether realists or positivists, seems now to be finally open.

The present book does not try to cover all the important issues concerning the foundations of quantum theory, but focuses on the Einstein, Podolsky, and Rosen paradox. Another book providing an adequate background for anyone with a graduate-level knowledge of quantum mechanics to follow the exciting new developments in this area has recently been written by D. Home[9] who has managed to give a fair and balanced presentation of different viewpoints, notwithstanding his own preference for the realist approach.

The book is organized as follows. The first chapter is historical and traces the development of the EPR paradox from the original argument of Einstein et al. (1935) to Wigner's probabilistic formulation of Bell's inequality (1970), dealing with Bohr's reply to Einstein et al. (1935), Schrödinger's (1935) and Furry's (1935) contributions, Bohm's simplification (1951), and Bell's inequality (1965) along the way. In the second chapter the principles of realism and locality are looked at, Bell's inequality is deduced from them, and other formulations of the paradox are considered. Other ways of discriminating between local realism and quantum mechanics are examined in the third chapter, in which the distinction is also made between weak inequalities, deduced from local realism alone and never violated experimentally, and strong inequalities, which are easier to violate because they depend on further assumptions regarding detection. The more general probabilistic treatment, which rests on a generalization of the deterministic reality criterion used by Einstein et al., is also dealt with. In the fourth chapter inequalities are deduced to distinguish local realism and quantum mechanics for pairs of neutral kaons. Proposed solutions, including variable probability of detection, are reviewed in the fifth chapter.

ACKNOWLEDGMENTS. We wish to express our gratitude to all those with whom we so often discussed the EPR paradox, Bell's theorem, and all that. We especially mention José R. Croca, Fasma Diele, Augusto Garuccio, Ramon Risco Delgado, and Caroline H. Thomson. The generous and friendly hospitality granted to one of us (A.A.) by the *Istituto per Ricerche di Matematica Applicata* of Bari and by its director Roberto Peluso during tenure of a grant awarded by the *Consiglio Nazionale delle Ricerche* is gratefully acknowledged.

REFERENCES

1. A. EINSTEIN, *The Born–Einstein Letters*, Macmillan, London (1971), pp. 170–171.
2. L. DE LA PENA and A. M. CETTO, *The Quantum Dice. An Introduction to Stochastic Electrodynamics*, Kluwer, Dordrecht (1996), p. 7.
3. J. S. BELL, letter to E. Santos. See: M. FERRERO and E. SANTOS, *Found. Phys.* **27**, 765–800 (1997) at p. 787.
4. R. P. FEYNMAN, R. B. LEIGHTON and M. SANDS, *The Feynman Lectures on Physics*, Addison-Wesley, Reading (1965), vol. III, p. 1–11.
5. J. R. CROCA, The limits of Heisenberg's Uncertainty Relations, in: *Causality and Locality in Modern Physics and Astronomy*, pp. 1–19, S. Jeffers *et al.* (eds.), Kluwer Academic Publ., Dordrecht (1997).
6. G. BINNIG and H. ROHRER, *Rev. Mod. Phys.* **59**, 615–625 (1987).
7. D. W. POHL, W. DENK and M. LANZ, *Appl. Phys. Lett.* **44**, 651–653 (1984).
8. H. HEINZELMANN and D. W. POHL, *Appl. Phys. A* **59**, 89–101 (1994).
9. D. HOME, *Conceptual Foundations of Quantum Physics*, Plenum Press, New York and London (1997).

Contents

Chapter 1

Early Formulations

The EPR paradox has evolved a good deal during the 35 years between the original paper of Einstein, Podolsky, and Rosen and Wigner's probabilistic version of Bell's inequality in 1970. It began as an experimentally impossible *gedankenexperiment*—with such physical idealizations as monochromatic plane waves and Dirac δ-functions—in which the fundamental *reality criterion* was formulated. Interference was not explicitly involved. Phase relations in configuration or tensor product spaces were given a central role by Furry and Schrödinger. Bohm simplified the paradox by reformulating it in terms of dichotomic observables regarding spin-$\frac{1}{2}$ particles, and addressed the experimental issue with Aharonov. Then in 1965 Bell deduced his celebrated inequality from the principles of local realism, and showed that it was grossly violated by quantum mechanical interference. Stronger inequalities and experiments to test them soon followed.

1.1. THE ORIGINAL EINSTEIN–PODOLSKY–ROSEN ARGUMENT

Position in quantum theory is represented by a linear Hermitian operator Q, which acts as a multiplication operator, multiplying the wave function by the independent variable q. So the eigenvalue equation

$$Qu(q; x) = qu(q; x)$$

is solved by an arbitrary real value of q and by the corresponding eigenfunction

$$u(q; x) = \delta(x - q) \tag{1}$$

which is a Dirac δ-function. The wave function (1) indicates a fixed position q. Since it can be expressed as

$$u(q; x) = \frac{1}{h} \int dp' \exp\left(ip' \frac{x - q}{\hbar}\right) \tag{2}$$

all values of momentum are equally likely.

1

The momentum operator is the same as the position operator in momentum space and has the form

$$P = -i\hbar \frac{\partial}{\partial x}$$

in position space. The eigenvalue equation

$$Pv(p; x) = pv(p; x)$$

is solved by arbitrary real values of p and by the corresponding eigenfunctions:

$$v(p; x) = \exp\left(\frac{ipx}{\hbar}\right) \tag{3}$$

The position density $|v(p; x)|^2$ of the wave function (1) is constant, so all positions are equally probable.

Whereas state $u(q; x)$ indicates a precise position, momentum is completely undetermined. The monochromatic plane wave $v(q; x)$, on the other hand, represents an exact momentum but an indefinite position. The incompatibility between position and momentum is expressed by the commutator

$$[Q, P] = i\hbar \tag{4}$$

not vanishing.

When the value of a measurable quantity can be predicted with certainty, it is natural to attribute a "reality" to the quantity and, hence, to the object in question. This attribution does not, however, necessarily depend on the actual performance of a measurement; it can be made, for instance, before a measurement and be eventually confirmed by measurement. Einstein, Podolsky, and Rosen[1] gave the following precise formulation of the idea:

> If, without in any way disturbing a system, we can predict with certainty (i.e., with probability equal to unity) the value of a physical quantity, then there exists an element of physical reality corresponding to this physical quantity.

Application of the reality criterion to the delta function (1) allows the attribution of an element of reality, a real position q, to the object in question. Another element of reality corresponding to momentum p can likewise be attributed to the same object by an application of the reality criterion to the wave function (3). The notations $u(q; x)$ and $v(p; x)$ indicate explicitly the objective physical properties q and p of the respective physical states. All this applies only at a particular time t_0. The wave function (1), for instance, would explode if allowed to evolve.

The elements of reality corresponding to P and Q cannot, using the EPR reality criterion, be attributed to a single object *at the same time*. So it is

FIGURE 1.1. The EPR paradox involves two correlated quantum systems α and β. In later versions of the paradox the systems are assumed to move in opposite directions and to be separated far away from one another.

consistent with the quantum formalism to assume that when P is measured and assumes a definite value, any previous element of reality corresponding to Q is undermined by the action quanta exchanged between the measuring apparatus and the observed atomic object. The attribution of elements of reality in such cases is pointless.

Things change significantly when *two* correlated quantum objects (α and β) are considered (Fig. 1.1). As is well known, the wave function can be assigned almost arbitrarily at a given instant, the ensuing evolution being given by the time-dependent Schrödinger equation. Possible fixed-time wave functions for the system $\varepsilon = \alpha + \beta$, made up of the subsystems α and β, are

$$\varphi(q_0; x_1, x_2) = \int dq' \, c(q') u_\alpha(q'; x_1) u_\beta(q_0 + q'; x_2) \tag{5}$$

$$\tilde{\varphi}(p_0; x_1, x_2) = \int dp' \, \tilde{c}(p') v_\alpha(p'; x_1) v_\beta(p_0 - p'; x_2) \tag{6}$$

where the notation for fixed-position and fixed-momentum wave functions is the same as before, the only change being the specification of the quantum object (α or β) to which they refer.

The meaning of $\varphi(q_0; x_1, x_2)$ is the usual one: for instance, a position measurement on α will give a result contained in the interval $q' \rightarrow q' + dq'$ with probability $|c(q')|^2 \, dq'$. However, if q' has been found for α from (5), it can be predicted that a position measurement for β will certainly give the result $q_0 + q'$. In other words, correlated position measurements made on α and on β will lead to results whose difference certainly equals q_0. It can then be concluded that to q_0 there corresponds an element of reality of $\alpha + \beta$. One can likewise conclude, from $\tilde{\varphi}$, that $\alpha + \beta$ has an element of reality corresponding to the *sum* p_0 of the momenta.

The simultaneous attribution of q_0 and p_0 to a pair of quantum objects is no longer excluded *a priori*—as with q and p for a single object—since the difference of positions and the sum of momenta are represented by commuting operators:

$$[Q_\alpha - Q_\beta, \, P_\alpha + P_\beta] = 0 \tag{7}$$

Einstein, Podolsky, and Rosen were able to find a wave function that allows the simultaneous attribution of the two elements of reality, namely

$$\psi(q_0, p_0; x_1, x_2) = \frac{1}{h} \int dp' \, \exp\left[\frac{i(x_1 - x_2 + q_0)p'}{\hbar}\right] \tag{8}$$

which can be written in the form (6) with $p_0 = 0$ and with $\tilde{c}(p') = 1/h$ (apart from a constant phase factor). It can also be written in the form (5) [with $c(q') = 1$] since the integral in (8) gives

$$\psi(q_0, 0; x_1, x_2) = \delta(x_1 - x_2 + q_0)$$
$$= \int dq' \, \delta(x_1 - q')\delta(q' - x_2 + q_0) \tag{9}$$

These results were used by EPR to establish that the quantum-mechanical description of physical reality cannot be complete. A theory is considered complete when *every element of physical reality attributable to a certain physical system in a given state has a counterpart in the mathematical description provided by the theory for that physical situation*. For example, the wave function (8) would provide a complete description of the pair (α, β) if no further elements of reality beyond q_0 and p_0 could be attributed to the pair. It can, however, be established that *individual positions and momenta* of α and β do possess a physical reality, leading thus to the conclusion that the quantum-mechanical description provided by (8) is not complete.

Consider a very large set E of similar pairs (α, β) all described by the wave function (8) (Fig. 1.2). It can then be predicted that measurements of the positions of α and β performed on individual pairs will give results that always satisfy the

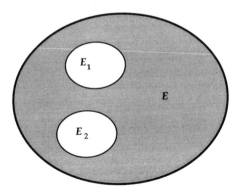

FIGURE 1.2. Set E is composed of (α, β) pairs. E_1 (E_2) is the subset of E on the α's of which a measurement of position (momentum) is performed. The EPR argument leads to the conclusion that elements of reality concerning position and momentum can be attributed simultaneously to all α's and β's of E.

relation $x_2 - x_1 = q_0$. Similarly, measurements of the momenta of α and β performed on (other) individual decay processes will give results p_1 and p_2 that always satisfy the relation $p_1 + p_2 = p_0 = 0$.

Now consider a subset E_1 of E not previously subjected to measurements and perform position measurements on all the α's of E_1; let $x_1', x_1'', \ldots$ denote the values obtained. It can then be predicted with certainty that simultaneous measurements of the position of the β's will give $x_1' - q_0$ for the first pair, $x_1'' - q_0$ for the second pair, and so on. One can then invoke the EPR criterion of physical reality and conclude that to the position of β there corresponds an element of physical reality. It is natural to conclude that this element of reality exists independently of measurement, for otherwise the measurement on α would have created β's element of reality at a distance and instantaneously. So to the position of β there corresponds an element of reality *for all the pairs of the full ensemble E.*

Likewise for momenta: consider a subset E_2 of E, perform a momentum measurement on all the α's of E_2, and let $p_1', p_1'', \ldots$ denote the results obtained. Since it can be predicted with certainty that subsequent measurements of the momentum of β will give $-p_1'$ for the first pair, $-p_1''$ for the second pair, and so on, it can also be concluded that to the momentum of β there corresponds an element of reality for all β's of E_2. Provided this element of reality was not created instantaneously by the distant measurement on α, one can then extend the foregoing conclusion to the whole of E. Clearly the choice of the system (α or β) on which measurements are performed is arbitrary: to the position and momentum of particle α, too, there therefore correspond simultaneous elements of reality in the whole ensemble E.

Individual positions and momenta are accordingly considered real before measurements, in an indirect sense, for all objects (α and β) of E; the sense being that there exists something in the physical reality of α and β that leads necessarily to preassigned results if a measurement of one or the other of the two observables is made.

Since the wave function (8) describes these quantities *a priori* as indeterminate, one must necessarily conclude, on the basis of the given definition of completeness, that the description of physical reality provided by (8) is not complete. In 1935 this was the essence of the EPR paradox, which was then only an argument to indicate the incompleteness of the existing theory.

1.2. BOHR'S REPLY: COMPLEMENTARITY

Bohr[2] claimed that his *complementarity*—"a new feature of natural philosophy"—dispensed with the paradox. This complementarity was supposed to imply a final renunciation of the classical ideal of causality and to require a radical revision of common attitudes to physical reality. Bohr contested not the EPR *demonstration* of incompleteness, whose validity he did not question, but its *premises*, which, he claimed, did not apply in the atomic domain.

A process is causal, for Bohr, if it takes place according to well-defined and identifiable rules, the most important being the laws of conservation of energy and momentum. The physicist who studies the phenomena of the atomic domain will naturally try to use his or her macroscopic preconceptions and to describe atomic processes as taking place both in space and time *and* according to energy and momentum conservation. However, he or she will discover that it is not possible to do so because quantum observables described by noncommuting operators cannot be measured simultaneously. The measurement of one of them in general destroys previous knowledge about others.

The character of complementarity can best be illustrated with space localization (position measurement) and causality implementation (momentum measurement). Space localization can be achieved by measuring position with infinite precision so that the indeterminacy Δx vanishes. A measurement yielding the position q turns the wave function into a δ-function $\delta(x - q)$; but such a δ-function again expresses a complete indefiniteness in momentum, as it can be written as a superposition of all possible plane waves with coefficients *of the same modulus*. Whatever may have been known about momentum before is thus lost. Hence, there can be no evidence about momentum conservation if nothing is known about momentum. Localization in space therefore leads to an abandonment of the causal description.

Symmetrically, in a different experiment, one could decide to implement the causal description by measuring momentum with infinite precision. The wave function would then become a plane wave. But this undermines the spatial description of the quantum phenomenon completely, since nothing is known about position. Causal description therefore precludes spatial description.

The two possibilities, space-time and causality, are hence mutually incompatible. Bohr concludes that in the atomic world it is, in principle, impossible to give a picture of quantum processes as developing causally in space and time, and that this element of irrationality is introduced in quantum physics by the finite value of Planck's constant. For these reasons it becomes necessary in his opinion to limit the interest of the physicist to the exclusive consideration of acts of observation.

Hence, no paradox exists when one considers two correlated systems described by the wave function (8). Consider two apparatuses Q_1 and P_1 (Q_2 and P_2) capable of performing, respectively, position and momentum measurements on the system $\alpha(\beta)$. If one chooses to use Q_1 and Q_2, the wave function (8) predicts that the results x_1 and x_2 will be precisely correlated: $x_1 - x_2 = q_0$. If, instead, one chooses to use P_1 and P_2, the wave function (8) predicts a precise correlation of the results p_1 and p_2: $p_1 + p_2 = 0$. The two apparatuses Q_1 and P_1

FIGURE 1.3. Bohr and Einstein walk down a boulevard in Brussels during the 1930 Solvay Conference (AIP Emilio Segrè Visual Archives).

are mutually incompatible: one can choose to employ either Q_1 or P_1, but never the two of them simultaneously; the same holds for Q_2 and P_2. From this point of view the EPR assumption about elements of reality becomes useless: now it can only lead to the conclusion that *an element of reality is associated with a concretely performed act of measurement*, for there is no other reality that one can speak of. In particular, the EPR conclusion that position and momentum correspond to two simultaneously existing elements of reality appears totally unjustified (Bohr says that it contains "an essential ambiguity"), because one can never perform simultaneous measurements of position and momentum.

Einstein, Podolsky, and Rosen[1] anticipated the possibility of such a refutation and found it unacceptable:

One could object to this conclusion on the grounds that our criterion of reality is not sufficiently restrictive. Indeed, one would not arrive at our conclusion if one insisted that one or more physical quantities can be regarded as simultaneous elements of reality only when they can be simultaneously measured or predicted. On this point of view, since either one or the other, but not both simultaneously, of the quantities P and Q can be predicted, they are not simultaneously real. This makes the reality of P and Q depend upon the process of measurement carried out on the first system, which does not disturb the second system in any way. No reasonable definition of reality could be expected to do this.

The foregoing considerations apply to any two noncommuting operators. Consider, for instance, a spin-$\frac{1}{2}$ particle and its spin-component operators S_x, S_y, and S_z. Any two of them do not commute, which expresses the incompatibility of the corresponding observables. Consider an electron in the spin state

$$u_+ = \begin{pmatrix} c_1 \\ c_2 \end{pmatrix}$$

with $|c_1|^2 + |c_2|^2 = 1$. If the observable associated with S_z is measured, there can be only the two outcomes

$$S_z = \pm \frac{\hbar}{2}$$

with respective probabilities $|c_1|^2$ and $|c_2|^2$. The spin part of the state vector after measurement becomes an eigenvector of S_z; that is,

$$\begin{pmatrix} 1 \\ 0 \end{pmatrix} \quad \text{or} \quad \begin{pmatrix} 0 \\ 1 \end{pmatrix} \tag{10}$$

In either case the S_x component is totally unknown because its eigenstates

$$\frac{1}{\sqrt{2}} \begin{pmatrix} 1 \\ 1 \end{pmatrix} \quad \text{and} \quad \frac{1}{\sqrt{2}} \begin{pmatrix} 1 \\ -1 \end{pmatrix}$$

can be written as superpositions of the states (10) with coefficients of equal modulus. Bohr would say that the implementation of the reality of S_z has made S_x completely undetermined—not just disturbed and unknown, but undetermined in some ontological sense. Likewise S_x can become known, but then S_z becomes necessarily completely unknown. One can thus say, with Bohr, that S_x and S_z are complementary aspects of reality; either S_x is real, or S_z is real, never both at the same time.

1.3. SCHRÖDINGER'S EXTENSION OF THE PARADOX

Schrödinger,[3] in 1935, considered a wave function [like (8)] satisfying the two eigenvalue equations

$$Q\psi(x_1, x_2) = q_0\psi(x_1, x_2)$$
$$P\psi(x_1, x_2) = p_0\psi(x_1, x_2) \tag{11}$$

where $Q = Q_\alpha - Q_\beta$ and $P = P_\alpha + P_\beta$ [the notation is the same as in the first section: see Eq. (7)], and showed that to every Hermitian operator $F(Q_\alpha, P_\alpha)$ of the first particle of an EPR pair there corresponds another Hermitian operator $G(Q_\beta, P_\beta)$ for the second particle, such that

$$[F(Q_\alpha, P_\alpha) - G(Q_\beta, P_\beta)]\psi(x_1, x_2) = 0 \tag{12}$$

Thus $\psi(x_1, x_2)$ is an eigenfunction of $F - G$ with eigenvalue zero. Measurements of F on α and of G on β must therefore give equal results if α and β are described by $\psi(x_1, x_2)$.

To prove Schrödinger's theorem, take the operator

$$F_{mn}(Q_\alpha, P_\alpha) = Q_\alpha^m P_\alpha^n + h.c. \tag{13}$$

with m and n integers, and assume that it corresponds to

$$G_{mn}(Q_\beta, P_\beta) = (Q_\beta + q_0)^m (p_0 - P_\beta)^n + h.c.$$

which can also be written

$$G_{mn}(Q_\beta, P_\beta) = (Q_\alpha - Q + q_0)^m (p_0 + P_\alpha - P)^n + h.c.$$

by definition of Q and P.

Hence when G_{mn} is applied to ψ, the factor $(p_0 + P_\alpha - P)^n$ becomes P_α^n because of (11). Since P_α obviously commutes with

$$Q_\beta + q_0 = Q_\alpha - Q + q_0$$

one can commute P_α^n to the extreme left of G_{mn}. On the right there remains a factor $Q_\alpha - Q + q_0$ which, applied to $\psi(x_1, x_2)$, gives Q_α^m. One thus obtains

$$G_{mn}\psi(x_1, x_2) = [P_\alpha^n Q_\alpha^m + h.c.]\psi(x_1, x_2)$$

The right-hand side coincides with $F_{mn}\psi(x_1, x_2)$ and thus (12) holds for the operator (13).

This can obviously be generalized to functions of the type

$$F(Q_\alpha, P_\alpha) = \sum_{mn} c_{mn} Q_\alpha^m P_\alpha^n + h.c. \tag{14}$$

where the c_{mn}'s are numerical coefficients. There is thus a wide class of operators, containing infinitely many terms, that satisfy Schrödinger's theorem: in practice, every analytic function $F(Q_\alpha, P_\alpha)$ can be developed as in (14) and must therefore satisfy Schrödinger's theorem.

In general, two such operators $F_1(Q_\alpha, P_\alpha)$ and $F_2(Q_\alpha, P_\alpha)$ [and their corresponding operators for β: $G_1(Q_\beta, P_\beta)$ and $G_2(Q_\beta, P_\beta)$] do not commute with one another. Still, since by Schrödinger's theorem the measurements of an F operator and of its corresponding G operator must in all cases give equal results, a measurement of F on α steers β into an eigenstate of G. Schrödinger concludes:

> It is rather discomforting that the theory should allow a system to be steered or piloted into one or the other type of state at the experimenter's mercy in spite of his having no access to it.

This problem will be reformulated for the dichotomic observables of spin-$\frac{1}{2}$ systems in Section 2.3.2.

1.4. FURRY'S HYPOTHESIS

If the states of particle α are described by vectors belonging to Hilbert space $\mathcal{H}^\alpha$, and those of particle β by vectors in $\mathcal{H}^\beta$, the states of the composite system $\varepsilon = (\alpha, \beta)$ will be represented by vectors in $\mathcal{H} \equiv \mathcal{H}^\alpha \otimes \mathcal{H}^\beta$. The tensor product $\mathcal{H}$ is not the same as the Cartesian product $\mathcal{H}^\alpha \times \mathcal{H}^\beta$, whose elements are all products $|\psi\rangle|\varphi\rangle$ of a vector $|\psi\rangle$ in $\mathcal{H}^\alpha$ and a vector $|\varphi\rangle$ in $\mathcal{H}^\beta$. Quantum mechanics is a linear theory whose states can be superposed to form new states; the sum of products will not always be another product. The linear space $\mathcal{H}$, on the other hand, includes all superpositions of products. If the sets $\{|\psi_i\rangle\} \subset \mathcal{H}^\alpha$ and $\{|\varphi_j\rangle\} \subset \mathcal{H}^\beta$ are complete and orthonormal, the set of products $|\psi_i\rangle|\varphi_j\rangle$ will form a basis in $\mathcal{H}$, so any linear combination

$$|\eta\rangle = \sum_{ij} c_{ij} |\psi_i\rangle|\varphi_j\rangle \tag{15}$$

with complex coefficients c_{ij} represents a possible description of ε. The coefficients $\{c_{ij}\}$, being totally arbitrary, will not necessarily be products of the form $c_{ij} = a_i b_j$. Indeed, if they were, $|\eta\rangle$ could be written as the product

$$\sum_{ij} a_i b_j |\psi_i\rangle|\varphi_j\rangle = \left(\sum_i a_i|\psi_i\rangle\right)\left(\sum_j b_j|\varphi_j\rangle\right) \in \mathcal{H}^\alpha \times \mathcal{H}^\beta \tag{16}$$

Only if a vector is a product can a state vector be assigned to either system. So the linearity of $\mathcal{H}$ means that the particles of ε may not have a state. Since the state vector is the only link between the quantum-mechanical formalism and micro-physical reality, *quantum theory does not ascribe any separate reality to the objects α and β whose complex (α, β) is described by a state vector that is not a* *product.* An embryo of the EPR paradox is already visible in this strange fact. More will be said about this in Section 2.3.1.

A general discussion of the EPR paradox was given by Furry[4,5] shortly after the publication of the EPR paper. His starting point was a theorem proved by von Neumann according to which the state vector (15) can always be given the 'polar' decomposition

$$|\eta\rangle = \sum_i c_i |\sigma_i\rangle |\tau_i\rangle \tag{17}$$

if the complete orthonormal sets $\{|\sigma_i\rangle\} \subset \mathcal{H}^\alpha$ and $\{|\tau_i\rangle\} \subset \mathcal{H}^\beta$ are suitably chosen. If $|\eta\rangle$ is a product, there will only be one term in the sum (17), and all the coefficients but one will vanish. Assume instead that two or more do not vanish, so it is impossible to write $|\eta\rangle$ as a product. Then construct the self-adjoint operators

$$A = \sum_i a_i |\sigma_i\rangle\langle\sigma_i| \qquad B = \sum_j b_j |\tau_j\rangle\langle\tau_j| \tag{18}$$

which, for simplicity, shall be assumed maximal. In other words, the a_i's are all different and so are the b_j's. The polar form correlates the two bases, and the maximality of A and B extends the correspondence to the sets of eigenvalues. If $C \equiv A \otimes B$ commutes with the Hamiltonian H of the composite system ε, time evolution will preserve the polar form

$$U(t)|\eta\rangle \equiv |\eta(t)\rangle = \sum_i c_i(t) |\sigma_i\rangle |\tau_i\rangle$$

with the same bases and will only affect the coefficients $c_i(t)$. Otherwise, to maintain the polar form, the bases will have to rotate as well, and the operators $A(t)$ and $B(t)$ will change correspondingly (even if they keep the same eigenvalues). Now provided C and H commute, if a measurement of A on the α particle yields the eigenvalue a_k, a measurement of the observable represented by B at any other time would necessarily yield b_k. Otherwise, if C and H do not commute, the two measurements have to be made at the same time to be correlated.

One can say that the eigenvalue b_k corresponding to the measured eigenvalue a_k was predetermined, since the two subsystems can be very far apart. Quantum theory has a precise way to describe the state of an object for which the result of a measurement is known *a priori*: It attributes as state vector the eigenvector corresponding to the known value of the considered physical quantity.

Given the polar form of $|\eta(t)\rangle$, the state of β must have been $|\tau_k\rangle$ before a_k was revealed. Given the predicted correlation of values for the observables A and B, one concludes that the state vector for (α, β) must have been $|\sigma_k\rangle|\tau_k\rangle$ in the first place. Repeating the previous argument for a statistical ensemble of pairs (α, β), one concludes that the state vectors actually were

$$
\begin{aligned}
|\eta_1\rangle &= |\sigma_1\rangle|\tau_1\rangle &&\text{in } |c_1|^2\% \text{ of the cases} \\
|\eta_2\rangle &= |\sigma_2\rangle|\tau_2\rangle &&\text{in } |c_2|^2\% \text{ of the cases} \\
&\;\;\vdots \\
|\eta_k\rangle &= |\sigma_k\rangle|\tau_k\rangle &&\text{in } |c_k|^2\% \text{ of the cases}
\end{aligned}
\tag{19}
$$

Schrödinger[6] suggested that the mixture (19) may eventually *become* a better description of the state once described by $|\eta\rangle$.

> ...[T]he paradox could be avoided by a very simple assumption, namely if the situation after separating were described by the expansion
>
> $$\Psi(x, y) = \sum_k a_k g_k(x) f_k(x),$$
>
> but with the additional statement that the knowledge of the phase relations between the complex constants a_k has been entirely lost in consequence of the process of separation. This would mean that not only the parts, but the whole system, would be in the situation of a mixture, not of a pure state.

Now (17) and (19) are obviously different *mathematical* descriptions of the ensemble. Could they nevertheless be equivalent descriptions for all practical purposes? Not even that, as Furry showed.

An elegant way to express the statistical distinguishability between superpositions and mixtures of products was found by Fortunato.[7] Consider the projection operator

$$
P_\eta = |\eta\rangle\langle\eta|
\tag{20}
$$

which is Hermitian and can be assumed to correspond to an observable. Its expectation value over the state (17) is obviously unity:

$$
\langle\eta|P_\eta|\eta\rangle = 1
\tag{21}
$$

The expectation value of the same operator over the mixture (19) is instead

$$
\langle P_\eta\rangle = \sum_j |c_j|^2 \langle\sigma_j\tau_j|P_\eta|\sigma_j\tau_j\rangle
$$

Since the expectation

$$
\langle\sigma_j\tau_j|P_\eta|\sigma_j\tau_j\rangle = |c_j|^2
$$

it follows that

$$
\langle P_\eta\rangle = \sum_j |c_j|^4
\tag{22}
$$

But the sum of the fourth powers of the moduli of numbers whose squared moduli add up to 1 will certainly be less than 1, if there are at least two numbers different from zero. The latter condition is, however, precisely that of having a state vector of the second type. Therefore, the observable corresponding to P_η has an expectation value equal to 1 [less than 1] over the state vector of the second type $|\eta\rangle$ [over the mixture of state vectors of the first type (19)].

1.5. BOHM'S SIMPLIFICATION OF THE PARADOX

Let a physical system ε be given (atom, molecule, ...) decaying into two spin-$\frac{1}{2}$ "particles" α and β. Let $u_\alpha(+)$ and $u_\alpha(-)$ be eigenvectors corresponding to the eigenvalues $+1$ and -1, respectively, of the Pauli matrix $\sigma_3(\alpha)$ representing the third component of the spin angular momentum for α, and let $u_\beta(+)$ and $u_\beta(-)$ be the corresponding eigenvectors of the Pauli matrix $\sigma_3(\beta)$ for β. The only factorizable spin states for (α, β) pairs one can construct with these four spinors are

$$u_\alpha(+)u_\beta(+), \quad u_\alpha(+)u_\beta(-), \quad u_\alpha(-)u_\beta(+), \quad u_\alpha(-)u_\beta(-) \qquad (23)$$

where the first one applies when the spin vectors of both particles α and β point in the positive z-direction, and so on.

There are physical situations in which the spin state vector for (α, β) must be the "singlet" state vector η_0 given by

$$\eta_0 = \frac{1}{\sqrt{2}}[u_\alpha(+)u_\beta(-) - u_\alpha(-)u_\beta(+)] \qquad (24)$$

Four important properties of η_0 are used in Bohm's[8] version of the EPR paradox:

(P1) It is not a factorizable state.
(P2) It predicts the result zero for a measurement of the total squared spin of particles α and β.
(P3) It is rotationally invariant.
(P4) It predicts opposite results for measurements of the components along $\hat{\mathbf{n}}$ of the spins of particles α and β, $\hat{\mathbf{n}}$ being an arbitrary unit vector.

Property (P1) is not difficult to prove, since the most general spin state for α is

$$U_\alpha = au_\alpha(+) + bu_\alpha(-) \qquad (|a|^2 + |b|^2 = 1) \qquad (25)$$

where a and b are constants. Similarly, the most general spin state for β is

$$U_\beta = cu_\beta(+) + du_\beta(-) \qquad (|c|^2 + |d|^2 = 1) \tag{26}$$

where c and d are other constants. Obviously the most general factorizable spin state for the combined system is

$$
\begin{aligned}
U_\alpha U_\beta &= acu_\alpha(+)u_\beta(+) + adu_\alpha(+)u_\beta(-) \\
&\quad + bcu_\alpha(-)u_\beta(+) + bdu_\alpha(-)u_\beta(-)
\end{aligned}
\tag{27}
$$

Now, since $u_\alpha(+)u_\beta(+)$ does not figure in η_0, $U_\alpha U_\beta$ can be equal to η_0 only if $ac = 0$. Thus $a = 0$, which implies that $u_\alpha(+)u_\beta(-)$ also disappears from $U_\alpha U_\beta$, and/or $c = 0$, which implies that $u_\alpha(-)u_\beta(+)$ disappears. It is therefore impossible, by any choice of U_α and U_β, to satisfy $\eta_0 = U_\alpha U_\beta$.

As for property (P2), it can be verified by introducing the total squared spin operator, defined, apart from a proportionality factor of $\hbar^2/4$, by

$$
\begin{aligned}
\Sigma^2 &= [\sigma_1(\alpha) + \sigma_1(\beta)]^2 + [\sigma_2(\alpha) + \sigma_2(\beta)]^2 + [\sigma_3(\alpha) + \sigma_3(\beta)]^2 \\
&= 6I_\alpha \otimes I_\beta + 2[\sigma_1(\alpha) \otimes \sigma_1(\beta) + \sigma_2(\alpha) \otimes \sigma_2(\beta) + \sigma_3(\alpha) \otimes \sigma_3(\beta)]
\end{aligned}
\tag{28}
$$

where I_α and I_β are the unit operators in the spin spaces of α and β, respectively. One can easily check that

$$\Sigma^2 \eta_0 = 0 \tag{29}$$

whence it follows that a measurement of the observable corresponding to Σ^2 on a pair described by the state η_0 will certainly give the result zero.

The third fundamental property of η_0 (rotational invariance) can be proved by introducing the new vectors $u_\alpha(n\pm)$ and $u_\beta(n\pm)$, which denote eigenvectors of $\boldsymbol{\sigma}(\alpha) \cdot \hat{\mathbf{n}}$ and $\boldsymbol{\sigma}(\beta) \cdot \hat{\mathbf{n}}$, respectively ($\hat{\mathbf{n}}$ being an arbitrary unit vector), and showing that η_0 transforms into

$$\eta_0 = \frac{1}{\sqrt{2}}[u_\alpha(n+)u_\beta(n-) - u_\alpha(n-)u_\beta(n+)] \tag{30}$$

which has the same structure as (24) with different states.

As for property (P4), it can be checked that η_0 is an eigenstate of the $\hat{\mathbf{n}}$-component of the total spin operator with eigenvalue zero; that is,

$$[\boldsymbol{\sigma}(\alpha) \cdot \hat{\mathbf{n}} + \boldsymbol{\sigma}(\beta) \cdot \hat{\mathbf{n}}]\eta_0 = 0 \tag{31}$$

From the physical interpretation of eigenvalue relations it then follows that measurements of the $\hat{\mathbf{n}}$ components of the spins of α and β must always give opposite results.

Another important state for the discussion of the EPR paradox is the "triplet" state, given by

$$\eta_1 = \frac{1}{\sqrt{2}}[u_\alpha(n+)u_\beta(n-) + u_\alpha(n-)u_\beta(n+)] \tag{32}$$

One can show that η_1 shares with η_0 properties (P1) and (P4), but not (P3); it is not rotationally invariant. Moreover, in place of (P2), η_1 has the following property: *Any measurement of the total squared spin of the two particles described by* η_1 *will give the result* $2\hbar^2$ *(spin 1).* One has

$$u_\alpha(+)u_\beta(-) = \frac{1}{\sqrt{2}}(\eta_0 + \eta_1)$$
$$u_\alpha(-)u_\beta(+) = \frac{1}{\sqrt{2}}(\eta_0 - \eta_1) \tag{33}$$

as can be proved simply by adding and subtracting (24) and (32). Furthermore, on invoking the quantum-mechanical interpretation of superpositions, one sees that *measurements of the total squared spin on a set of* (α, β) *pairs described as a mixture of the factorizable state vectors (33) will produce with equal probability the results* 0 *and* $2\hbar^2$.

This large observable difference between an ensemble which is an arbitrary mixture of the states (33) and an ensemble whose elements are all described by η_0 is the basis of the Bohm formulation of the paradox. This formulation has several important advantages over the original one. First, it deals with dichotomic observables and therefore allows for a sharper definition of the results. Second, it allows the introduction of time, which figures only in the space-dependent part of the wave function, while the spin part is in most cases time independent. Therefore the singlet state is, as it were, stable, while the original wave function (8)–(9) introduced by Einstein, Podolsky and Rosen holds only at a particular time and blows up immediately after. The third advantage of the Bohm formulation is that it allows one to deal with clearly separated objects (in space), while the wave function (8), based on plane waves, described the two correlated objects as present, with constant probability, in all points of space.

In order to establish the EPR paradox in conceptually clear conditions, consider only (α, β) pairs with the wave function

$$\psi(x_1, x_2) = \eta_0 \psi_\alpha(x_1)\psi_\beta(x_2) \tag{34}$$

where η_0 is the singlet state (24) and $\psi_\alpha(x_1)$, $\psi_\beta(x_2)$ are the space parts—assumed factorizable—of the wave functions for α and β, respectively. Suppose furthermore that $\psi_\alpha(x_1)$ is a Gaussian function with modulus appreciably different from zero only in a region R_1 of width Δ_1 centered around the point x_{10}. Similarly, let $\psi_\beta(x_2)$ be a Gaussian function localized in the region R_2 of width Δ_2 centered around x_{20}. The condition

$$|x_{20} - x_{10}| \gg \Delta_1, \Delta_2 \qquad (35)$$

will be considered sufficient for the separability of systems α and β. If particles α and β are supposed to move to the left and to the right, respectively, so that the distance between the centers of the two wave packets increases linearly with time, it is not difficult to show that Schrödinger's equation allows situations in which the condition for separability does not deteriorate with time. One can thus say that α and β are located within two small regions R_1 and R_2, respectively, very far from one another so that all mutual interactions (gravitational, electromagnetic, strong and weak) are known to be very small. In such conditions it is natural to conclude that a measurement performed on α does not give rise to any modification of the physical properties of β, and *vice versa*. The presence of η_0 in (34) leads instead to paradoxical conclusions.

The EPR reasoning is as follows. Consider a large set E of (α, β) pairs in the state (34). Measure $\sigma_3(\alpha)$ at time t_0 on all α's of a subset E_1 of E. If $+1$ (-1) is found, a future ($t > t_0$) measurement of $\sigma_3(\beta)$ will certainly give -1 ($+1$). Using the EPR reality criterion, one can assign to the β's of E_1 an element of reality λ_1 (λ_2) fixing *a priori* the result -1 ($+1$) of the $\sigma_3(\beta)$ measurement.

But quantum mechanics treats an object β with predetermined value of $\sigma_3(\beta)$ by assigning it the state $u_\beta(-)$ [$u_\beta(+)$]; this is the completeness assumption. The strict correlation (P4), applied to the z-axis, implies then, even for $t < t_0$, that the ensemble E_1 had to be described in spin space by the mixture (33). Excluding that λ_1 (λ_2) is created at a distance by the measurement of $\sigma_3(\alpha)$, it must be concluded that λ_1 (λ_2) actually belongs to all β's of E. Applying completeness once more, one again concludes that the mixture (33) applies to all pairs of E. But this contradicts the description (34) at the empirical level, as was shown. One therefore reaches an absurd conclusion (EPR paradox).

1.6. BOHM–AHARONOV AND THE EXPERIMENTAL ISSUE

In 1936 it was clear that the EPR paradox had brought out the existence of a striking

> ... disagreement between the results of quantum-mechanical calculations and those to be expected on the assumption that a system once freed from dynamical interference can be regarded as possessing independently real properties.[4]

This conclusion is very difficult to accept for some people and can lead to the idea that something must be wrong with the existing quantum theory. Even Einstein entertained such a view:

...Einstein has (in private communication) actually proposed such an idea; namely, that the current formulation of the many-body problem in quantum mechanics may break down when particles are far enough apart.[9]

The first examination of a possible breakdown of quantum theory (with immediately negative conclusions) was made by Bohm and Aharonov[9] in 1957. They considered the annihilation of a positron–electron pair into two energetic photons (gamma rays) and showed that the quantum state produced is

$$|0^-\rangle = \frac{|x_\alpha\rangle |y_\beta\rangle - |y_\alpha\rangle |x_\beta\rangle}{\sqrt{2}} \qquad (36)$$

that is, the zero-angular-momentum negative-parity state, where x and y denote the directions of linear polarization of photon α and photon β. The latter state, like the singlet state of two spin-$\frac{1}{2}$ particles considered earlier, is rotationally invariant. In practice this means that each photon is always found in a state of linear polarization orthogonal to that of the other, regardless of the choice of axes with respect to which the state of polarization is expressed. Bohm and Aharonov calculated the ratio $R = \Gamma_1/\Gamma_2$, where Γ_1 is the rate of double scattering of the two photons through a fixed angle θ, when the planes π_1 and π_2 formed by the lines of motion of the first and second photons (after scattering) with their common original direction of motion are *perpendicular*; Γ_2 is the same rate when the planes π_1 and π_2 are *parallel*.

The value of R predicted by the $|0^-\rangle$ state is

$$R = \frac{(\gamma - 2\sin^2\theta)^2 + \gamma^2}{2\gamma(\gamma - 2\sin^2\theta)} \qquad (37)$$

where

$$\gamma = \frac{k_0}{k} + \frac{k}{k_0} \qquad (38)$$

Here k_0 is the wave number of the incident photon, and k is that of the final photon. Bohm and Aharonov considered an angle of 82° for which the ratio k_0/k can easily be calculated from Compton scattering kinematics, and obtained $R = 2.85$. This figure could not be compared directly with the experiment of Wu and Shaknov[10] because photons were detected with an angular spread around the ideal value of 82°. Instead, to such a concrete situation the prediction $R = 2.00$ applied, obtained with a suitable angular average of (37), which agrees very well with the experimental result $R = 2.04 \pm 0.08$.

Bohm and Aharonov also showed that *the hypothesis of a breakdown of the* $|0^-\rangle$ *state vector with the increasing distance between the two photons and of its substitution with mixtures of factorizable vectors led necessarily to considerably smaller values of R*, always satisfying $R \leq 1.5$. These results show that Wu and Shaknov's experiment is explained adequately by existing quantum theory, which implies distant correlations of the type leading to the EPR paradox, but not by any approach, such as Furry's hypothesis, implying a simple-minded breakdown of the quantum theory that could avoid the paradox.

It would, however, not be correct to conclude that this experimental evidence constitutes an argument against local realism, since there are well-known local models capable of reproducing quantum-mechanical predictions for the experiments on double rescattering following electron–positron annihilation into two gamma rays. One such model was proposed by Kasday.[11] Suppose there are two "hidden" vectors $\hat{\boldsymbol{\lambda}}_1$ and $\hat{\boldsymbol{\lambda}}_2$ associated with photon α and photon β, respectively, and let the photons ultimately scatter in the directions of these vectors. Then simply give $\hat{\boldsymbol{\lambda}}_1$ and $\hat{\boldsymbol{\lambda}}_2$ the same probability distribution $\rho(\hat{\boldsymbol{\lambda}}_1, \hat{\boldsymbol{\lambda}}_2)$ as that of the momenta $\mathbf{k}_1$ and $\mathbf{k}_2$ of the scattered photons:

$$\rho(\hat{\boldsymbol{\lambda}}_1, \hat{\boldsymbol{\lambda}}_2) = H(\hat{\boldsymbol{\lambda}}_1, \hat{\boldsymbol{\lambda}}_2)$$

where $H(\mathbf{k}_1, \mathbf{k}_2)$ is the probability distribution of the momenta $\mathbf{k}_1$ and $\mathbf{k}_2$ of the scattered photons, as predicted by quantum theory. The assumption is clearly that the photons have "decided in advance," at the time of annihilation, in which direction they ultimately scatter. The model is local; changing the position of "detector 1" does not affect parameter $\hat{\boldsymbol{\lambda}}_2$, for example, and therefore it does not change the response of "detector 2." Furthermore, the model reproduces the results of all measurements that can be made on the scattered photons.

Given the conclusive evidence found by Bohm and Aharonov against a simple-minded breakdown of quantum theory, it is surprising that several authors rediscovered their idea long after it had been discarded by its proponents. Thus, Jauch[12] developed an ambitious approach to quantum theory based on an "algebra of propositions," where he defined as "mixtures of the 2nd kind" the quantum-mechanical description of EPR pairs based on nonfactorizable state vectors, that is, the states that lead to the EPR paradox, and concluded that "mixtures of the 2nd kind do not exist."

1.7. EPR CORRELATIONS FOR PAIRS OF NEUTRAL KAONS

A very interesting two-state system belonging to the domain of elementary particles is the meson called neutral kaon: symbol K^0, mass about $494\,\mathrm{MeV}/c^2$, spin zero. Neutral kaon pairs can be generated in the decay of ϕ mesons at rest, e.g., produced in electron–positron collisions:

$$e^+ + e^- \rightarrow \phi \rightarrow K^0 + \overline{K}^0$$

Given the quantum numbers of ϕ and the usual conservation laws of angular momentum J, parity P, and charge conjugation C in its (strong) decay, the final neutral kaons are described quantum mechanically by the $J^{PC} = 1^{--}$ state vector

$$|\Psi\rangle = \frac{1}{\sqrt{2}}\{|K^0\rangle_a|\overline{K}^0\rangle_b - |\overline{K}^0\rangle_a|K^0\rangle_b\} = \frac{1}{\sqrt{2}}\{|K_S\rangle_a|K_L\rangle_b - |K_L\rangle_a|K_S\rangle_b\} \quad (39)$$

where a (left) and b (right) denote the opposite directions of motion of the kaons, $|K^0\rangle$ and $|\overline{K}^0\rangle$ are the well-known state vectors for strangeness $+1$ and -1, respectively, and $|K_S\rangle$ and $|K_L\rangle$ are the state vectors for short- and long-lived kaons, respectively. The latter are eigenvectors of the CP operator. Kaon pairs in the state (39) are also known to be produced in $\bar{p} + p$ annihilations at rest from the prevalent $^3S^1$ state, namely in the reaction $\bar{p} + p \rightarrow K^0 + \overline{K}^0$.

Time evolution is obtained by applying a time evolution operator, which is taken to be nonunitary, in order to reproduce the exponential decrease of kaonic wave functions due to spontaneous disintegrations. One has

$$|K_S(t)\rangle = |K_S\rangle\exp(-\alpha_S t), \qquad |K_L(t)\rangle = |K_L\rangle\exp(-\alpha_L t) \quad (40)$$

where t is the particle proper time and

$$\alpha_S = \tfrac{1}{2}\gamma_S + im_S, \qquad \alpha_L = \tfrac{1}{2}\gamma_L + im_L \quad (41)$$

The small effect of CP nonconservation (unknown in 1961) is neglected in (40) and the $CP = \pm 1$ eigenstates are identified with the short- and long-living kaon, respectively. In Eq. (41), γ_S and m_S (γ_L and m_L) denote the decay rate and mass, respectively, of the S (L) meson. Units $\hbar = c = 1$ have been adopted. The time evolution operator for the state vector (39) is the product of the time evolutions for the individual kaons, so at proper times t_a and t_b one has

$$|\psi(t_a, t_b)\rangle = \frac{1}{\sqrt{2}}\{|K_S\rangle_a|K_L\rangle_b \exp(-\alpha_S t_a - \alpha_L t_b) - |K_L\rangle_a|K_S\rangle_b \exp(-\alpha_L t_a - \alpha_S t_b)\}$$

$$(42)$$

The difference between the two exponential factors in (42) generates $K^0 K^0$ and $\overline{K}^0\overline{K}^0$ components, which are absent at time zero, in (39). For example, the probability of a double $\overline{K}^0$ observation at times t_a and t_b is given by

$$P^{QM}[\overline{K}(t_a); \overline{K}(t_b)] = \tfrac{1}{8}[E_S(t_a)E_L(t_b) + E_L(t_a)E_S(t_b)$$
$$- 2\sqrt{E_S(t_a)E_L(t_b)E_L(t_a)E_S(t_b)}\, \cos \Delta m(t_a - t_b)] \quad (43)$$

where

$$E_S = e^{-\gamma_S t}, \qquad E_L = e^{-\gamma_L t} \quad (44)$$

and $\Delta m = m_L - m_S$ is the $K_L - K_S$ mass difference.

A remarkable consequence of the state vector (42) was observed by Lipkin[13]: the observation of a particular decay mode, i.e., $\pi^+\pi^-$, for one kaon implies that the other kaon is forbidden to decay, *at equal proper time*, in this mode. In fact the relevant transition matrix element for $t_a = t_b = t$ is $\langle D_a D_b | T | \psi(t, t) \rangle$, if

$$|D_x\rangle = |\pi^+ \pi^-\rangle_x \qquad (x = a, b) \tag{45}$$

A straightforward calculation based on (42) and using the transition matrix T gives

$$\langle D_a D_b | T | \psi(t, t) \rangle = \frac{1}{\sqrt{2}} \{ \langle D | T | K_S \rangle_a \langle D | T | K_L \rangle_b - \langle D | T | K_L \rangle_a \langle D | T | K_S \rangle_b \} e^{-(\alpha_S + \alpha_L)t}$$

$$= 0 \tag{46}$$

The vanishing of (46) is due to space isotropy because if all directions in space are equivalent, one must have

$$\langle D | T | K_S \rangle_a = \langle D | T | K_S \rangle_b \qquad \text{and} \qquad \langle D | T | K_L \rangle_a = \langle D | T | K_L \rangle_b \tag{47}$$

One can show that the previous selection rule applies even if *CP* violation is taken into account. The main technical reason behind this generalization is that *CP* violation modifies the state vector (42) only in the multiplying factor, while the term in braces remains the same. The decay of both kaons into two neutral pions is similarly forbidden.

The essential physical point of these phenomena can be summarized as follows: The observation of a particular decay mode $|D\rangle$ at time t in the positive (a) direction imposes constraints on the kaon beam observed in coincidence in the negative (b) direction. These constraints force $|K\rangle_b$ to be a definite linear combination of $|K_S\rangle$ and $|K_L\rangle$. It is just that linear combination which at the same t is forbidden to decay into the mode $|D\rangle$.

Lipkin described this situation as "the Einstein–Podolsky–Rosen effect." This terminology is not fully adequate, however, because the mere anticorrelation at equal proper times has nothing paradoxical and can be reproduced by local realistic models. The whole matter will be taken up in detail in Chapter 4.

1.8. BELL'S INEQUALITY

Again consider an ensemble formed by a very large number N of decays $\varepsilon \to \alpha + \beta$ and suppose that the observer O_α measures on α the dichotomic observable $A(a)$, while in a distant region of space a second observer measures on β another dichotomic observable $B(b)$.

The observables $A(a)$ and $B(b)$ have been taken to depend on the arguments a and b, respectively, which are assumed to be experimental parameters, fixed in the structure of the apparatuses in any given experiment, but possibly variable over different experiments. Examples of such dichotomic observables are those represented by the spin matrices $\boldsymbol{\sigma}(\alpha) \cdot \hat{\mathbf{a}}$ and $\boldsymbol{\sigma}(\beta) \cdot \hat{\mathbf{b}}$, where the experimental parameters are the unit vectors $\hat{\mathbf{a}}$ and $\hat{\mathbf{b}}$. They could be fixed experimentally, for example, by the directions of the inhomogeneous magnetic fields of two Stern–Gerlach apparatuses.

In any event, when measurements of such observables are made on all the N pairs of the given ensemble, O_α will obtain a set of results $\{A_1, A_2, \ldots, A_N\}$ while O_β will collect a similar set $\{B_1, B_2, \ldots, B_N\}$, all relative to fixed values of the parameters a and b. The results of the two sets are correlated in the sense that A_1 and B_1 pertain to particles α and β, respectively, arising from the first decay; A_2 and B_2 are similarly associated with the second decay, and so on. By definition, these results in every case are equal to ± 1.

The correlation function $P(a, b)$ of the results A_i and B_i is defined as the average product of the results obtained by O_α and O_β:

$$P(a, b) = \frac{1}{N} \sum_{i=1}^{N} A_i B_i \tag{48}$$

Since every product $A_i B_i$ is equal to ± 1, it follows that

$$-1 \leq P(a, b) \leq +1 \tag{49}$$

The quantum-mechanical correlation function in the case of singlet state η_0 given by (24) is equal to

$$P(\hat{\mathbf{a}}, \hat{\mathbf{b}}) = \langle \eta_0 | \boldsymbol{\sigma}(\alpha) \cdot \hat{\mathbf{a}} \otimes \boldsymbol{\sigma}(\beta) \cdot \hat{\mathbf{b}} | \eta_0 \rangle = -\hat{\mathbf{a}} \cdot \hat{\mathbf{b}} \tag{50}$$

This result is simple and elegant, but incompatible with local realism, as shall soon be seen.

Now define the quantity

$$\Delta = P(\hat{\mathbf{a}}, \hat{\mathbf{b}}) - P(\hat{\mathbf{a}}, \hat{\mathbf{b}}') + P(\hat{\mathbf{a}}', \hat{\mathbf{b}}) - P(\hat{\mathbf{a}}', \hat{\mathbf{b}}') \tag{51}$$

Consider two orthogonal unit vectors $\hat{\mathbf{a}}$ and $\hat{\mathbf{a}}'$ associated with particle α and two orthogonal unit vectors $\hat{\mathbf{b}}$ and $\hat{\mathbf{b}}'$ associated with particle β, and suppose that their orientation is such that they can be reached by clockwise rotations of $\pi/4$ in the order $\hat{\mathbf{a}}$, $\hat{\mathbf{b}}$, $\hat{\mathbf{a}}'$, $\hat{\mathbf{b}}'$ (Fig. 1.4). One can then easily see that substituting (50) in (51) leads to

$$|\Delta| = |\hat{\mathbf{a}} \cdot \hat{\mathbf{b}} - \hat{\mathbf{a}} \cdot \hat{\mathbf{b}}' + \hat{\mathbf{a}}' \cdot \hat{\mathbf{b}} + \hat{\mathbf{a}}' \cdot \hat{\mathbf{b}}'| = 2\sqrt{2}$$

It can, moreover, be shown that $2\sqrt{2}$ is the maximum value of $|\Delta|$ for all possible orientations of $\hat{\mathbf{a}}$, $\hat{\mathbf{a}}'$, $\hat{\mathbf{b}}$, $\hat{\mathbf{b}}'$. This result is of great interest, because local realism allows Δ to have a maximum value of 2. The result $\Delta \leq 2$ is *Bell's inequality*. It has been called "the most profound discovery of science."[14]

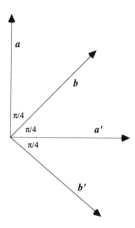

FIGURE 1.4. The spin components of the spin-$\frac{1}{2}$ particles (α, β) are jointly measured along directions (1) ($\hat{\mathbf{a}}$, $\hat{\mathbf{b}}$), (2) ($\hat{\mathbf{a}}$, $\hat{\mathbf{b}}'$), (3) ($\hat{\mathbf{a}}'$, $\hat{\mathbf{b}}$), (4) ($\hat{\mathbf{a}}'$, $\hat{\mathbf{b}}'$), in four different subsets of (α, β) pairs. The quantum-mechanical prediction (50) for singlet state leads to the maximum violation of Bell's inequality for the relative positions of the unit vectors $\hat{\mathbf{a}}$, $\hat{\mathbf{a}}'$, $\hat{\mathbf{b}}$, and $\hat{\mathbf{b}}'$ shown.

In a theory developed according to the EPR reality criterion, there are elements of reality λ that fix all observables. In general they can be expected to vary, with density $\rho(\lambda)$, over a set Λ. Of course,

$$\int_{\Lambda} d\lambda \, \rho(\lambda) = 1 \tag{52}$$

The role of the new variable λ is that of fixing the values of the dichotomic observables, for example,

$$\boldsymbol{\sigma}(\alpha) \cdot \hat{\mathbf{a}} \to A(a, \lambda), \qquad \boldsymbol{\sigma}(\beta) \cdot \hat{\mathbf{b}} \to B(b, \lambda) \tag{53}$$

where the discontinuous functions $A(a, \lambda)$ and $B(b, \lambda)$ can assume only the values ± 1. The correlation function as defined in (48) (average product of the two observables) can obviously be written

$$P(a, b) = \int d\lambda \, \rho(\lambda) A(a, \lambda) B(b, \lambda) \tag{54}$$

This is a local expression, in the sense that neither A depends on b nor B on a. It is easy to show that

$$|P(a, b) - P(a, b')| \leq \int d\lambda \, \rho(\lambda) |B(b, \lambda) - B(b', \lambda)| \tag{55}$$

since $|A(a, \lambda)| = 1$, and that

$$|P(a', b) + P(a', b')| \leq \int d\lambda \ \rho(\lambda)|B(b, \lambda) + B(b', \lambda)| \tag{56}$$

By adding (55) and (56) and using

$$|B(b, \lambda) - B(b', \lambda)| + |B(b, \lambda) + B(b', \lambda)| = 2$$

which is a consequence of $|B(b, \lambda)| = |B(b', \lambda)| = 1$, one obtains from (52) the inequality

$$|\Delta| \leq |P(a, b) - P(a, b')| + |P(a', b) + P(a', b')| \leq 2 \tag{57}$$

This is Bell's inequality in its 1970 form,[15] which is more general than its original 1965 form.[16]

1.9. ADDITIONAL ASSUMPTION AND STRONG INEQUALITIES

A practical way of testing experimentally the validity in nature of Bell's inequality could be the following: A source is built in such a way that the decays $\varepsilon \to \alpha + \beta$ lead to the emission of the pair only when the object α (β) flies to the left (to the right) where a two-channel analyzing apparatus can transmit it or reflect it at $90°$ depending on its physical properties. The dichotomic choice forced in this way upon the atomic objects can then be used for defining Bell's dichotomic observables, by saying that $A(a) = \pm 1$ [$B(b) = \pm 1$], depending on the channel, transmission or reflection, chosen by the object α [β].

In 1969, Clauser, Horne, Shimony, and Holt[17] (CHSH) suggested the use of pairs of optical photons emitted by atomic cascades. For such photons they assumed that the *binary choice was between transmission and absorption in a polarizer.* For every choice of the polarizer's orientations a and b, they introduced four probabilities, $T(a\pm, b\pm)$, where, for instance, $T(a+, b-)$ is the probability that observer O_α finds $A(a) = +1$ (photon α transmitted through polarizer with axis a) and that O_β finds $B(b) = -1$ (photon β absorbed by polarizer with axis b) (Fig. 1.5). The correlation function can then be written

$$P(a, b) = T(a+, b+) - T(a+, b-) - T(a-, b+) + T(a-, b-) \tag{58}$$

since the product of the results obtained by O_α and O_β is $+1$ [-1] in the cases of $T(a+, b+)$ and $T(a-, b-)$ [in the cases of $T(a-, b+)$ and $T(a+, b-)$]. Of course,

$$T(a+, b+) + T(a+, b-) + T(a-, b+) + T(a-, b-) = 1 \tag{59}$$

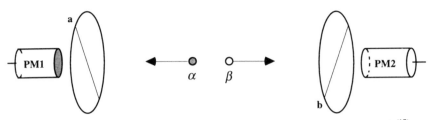

FIGURE 1.5. The EPR experiment proposed in 1969 by Clauser, Horne, Shimony, and Holt.[17] An atomic source emits a pair of photons α and β that travel in opposite directions and impinge on two polarizers with optical axes $\hat{a}$ and $\hat{b}$. If they get through the polarizers, they enter two photomultipliers **PM1** and **PM2**.

Considering further the case in which the second polarizer has been removed (denoted by ∞), one will obviously get

$$T(a+, b+) + T(a+, b-) = T(a+, \infty) \tag{60}$$

where $T(a+, \infty)$ is the joint probability for photon α to be transmitted through the polarizer with axis a and photon β to be transmitted through the vacuum (transmission through the vacuum, in the absence of a polarizer, is certain to take place); $T(\infty, b+)$ is defined analogously. If, instead, the first polarizer has been removed, one similarly gets

$$T(a+, b+) + T(a-, b+) = T(\infty, b+) \tag{61}$$

Finally, if both polarizers have been removed, both photons will certainly be transmitted, so

$$T(\infty, \infty) = 1 \tag{62}$$

Using Eqs. (59) to (61), it is a simple matter to show that the correlation function can be written as

$$P(a, b) = 4T(a+, b+) - 2T(a+, \infty) - 2T(\infty, b+) + 1 \tag{63}$$

In the latter expression only cases of double transmission appear, which are nearer to experimental observation, since it is impossible to detect the absorption of a photon in a polarizer. However at this point one must face a very important problem: Can one really measure the right-hand side of Eq. (63) with an error, say, of a few percent? Obviously, the only way to know that a photon has been transmitted through a polarizer is to detect its presence beyond the instrument, but the problem is that photon detectors have a low efficiency, typically 10% to 20%. This means that one cannot really measure a double-transmission probability, but only a joint probability for double transmission *and* double detection of the two photons. This is not what figures in Eq. (63)!

One could attempt to redefine the correlation function by using only the measurable joint probabilities for transmission and detection. This can certainly be done, but the trouble is that the values of $P(a, b)$ turn out to be of the order of 10^{-2}, which is far too small to lead to a violation of Bell's inequality.

This problem has traditionally been "solved" by means of *ad hoc* assumptions concerning the nature of the transmission–detection process. The additional assumption made by CHSH is the following:

> Given that a pair of photons emerges from two regions of space where two polarizers can be located, the probability of their joint detection from two photomultipliers is independent of the presence and of the orientation of the polarizers.

Calling ω_0 the double-detection probability dealt with in the previous assumption and denoting by ω the joint probability for transmission *and* detection of both photons, one can translate the previous assumption into the following relations:

$$
\begin{aligned}
\omega(a, b) &= \omega_0 T(a+, b+) \\
\omega(a, \infty) &= \omega_0 T(a+, \infty) \\
\omega(\infty, b) &= \omega_0 T(\infty, b+) \\
\omega(\infty, \infty) &= \omega_0 T(\infty, \infty)
\end{aligned}
\tag{64}
$$

where $\omega(a, b)$ is the joint probability in the case of polarizers with orientations a and b, $\omega(a, \infty)$ is the joint probability with the second polarizer removed and the first one oriented along a, and so on.

The rates of double detections depend, of course, on the number N_0 of photon pairs entering, per second, into the right solid angles defined by the optical apparatuses. Using R to denote rates, one has

$$
\begin{aligned}
R(a, b) &= N_0 \omega(a, b) \\
R(a, \infty) &= N_0 \omega(a, \infty) \\
R(\infty, b) &= N_0 \omega(\infty, b) \\
R_0 &= N_0 \omega_0
\end{aligned}
\tag{65}
$$

where $R(\infty, \infty)$ has been called R_0 and the meaning of the new symbols is obvious. If one obtains the T functions from the relations (64) and (65) and substitutes them in Eq. (63), one gets

$$
P(a, b) = 4\frac{R(a, b)}{R_0} - 2\frac{R(a, \infty)}{R_0} - 2\frac{R(\infty, b)}{R_0} + 1
\tag{66}
$$

Only coincidence rates enter into Eq. (66). By virtue of the CHSH additional assumption the correlation function has therefore become measurable!

Equation (66) allows Bell's inequality to be transformed into a directly measurable expression. In fact, inequality (57) is equivalent to

$$
-2 \le P(a, b) - P(a, b') + P(a', b) + P(a', b') \le +2
\tag{67}
$$

Substituting into the previous inequalities the expressions of the type (66) for the four correlation functions, one obtains

$$-1 \leq \frac{R(a, b)}{R_0} - \frac{R(a, b')}{R_0} + \frac{R(a', b)}{R_0} + \frac{R(a', b')}{R_0} - \frac{R(a', \infty)}{R_0} - \frac{R(\infty, b)}{R_0} \leq 0 \quad (68)$$

Only coincidence rates enter the previous inequalities, which can therefore be checked experimentally.

A useful simplification is obtained if two qualitative predictions of quantum theory, which have nothing paradoxical and which can be checked directly in experiments, are accepted:

1. The prediction that $R_1 \equiv R(a', \infty)$ does not depend on a' and that $R_2 \equiv R(\infty, b)$ does not depend on b.

2. The prediction that every R function should depend only on the relative angle between the polarizers' axes. For example, $R(a, b) = R(a - b)$.

Adopting these simplifications, one gets

$$-1 \leq \frac{R(a - b)}{R_0} - \frac{R(a - b')}{R_0} + \frac{R(a' - b)}{R_0} + \frac{R(a' - b')}{R_0} - \frac{R_1}{R_0} - \frac{R_2}{R_0} \leq 0 \quad (69)$$

The axes of the polarizers can be chosen in such a way that

$$a - b = a' - b = a' - b' = \phi, \qquad a - b' = 3\phi \quad (70)$$

Then from Eq. (69) it follows that

$$-1 \leq \frac{3R(\phi)}{R_0} - \frac{R(3\phi)}{R_0} - \frac{R_1 + R_2}{R_0} \leq 0 \quad (71)$$

Considering the previous inequalities for $\phi = 22\frac{1}{2}°$ and for $\phi = 67\frac{1}{2}°$, for which the maximal quantum-mechanical violation of (71) takes place, one can easily obtain the so-called Freedman inequality:

$$\delta = \left| \frac{R(22\frac{1}{2}°)}{R_0} - \frac{R(67\frac{1}{2}°)}{R_0} \right| - \frac{1}{4} \leq 0 \quad (72)$$

which does not involve R_1 or R_2.

We repeat that all the new results deduced, starting from Eq. (66) and ending with inequality (72), *have become possible only because the CHSH assumption has been made*. It is therefore not correct to confuse Bell's original inequality with the *much stronger* inequalities now deduced. In the future the following definitions will be adopted:

1. *Weak inequality*: An inequality deduced from the sole assumption of local realism and violated by quantum mechanics in the case of nearly perfect instruments.

2. *Strong inequality*: An inequality deduced from local realism and from *ad hoc* additional assumptions, such as the CHSH hypothesis, or other hypotheses to be seen later, and violated by quantum mechanics in the case of real instruments.

The difference between these two types of inequalities will be discussed at length in Chapter 3.

1.10. WIGNER'S PROOF OF BELL'S INEQUALITY

Wigner's 1970 proof[18] of Bell's inequality will be reviewed in a simpler and more general form than the original one. Of course, the basic ideas are strictly the same.

Wigner made two basic assumptions. The first one was that the results of all conceivable measurements are predetermined (even in the case of incompatible observables). This realistic standpoint does not contradict Heisenberg's relations because the latter can be taken simply to mean that a concrete measurement made on a given object modifies the preset values of other observables of that object, not compatible with the measured one. But, *before the action of the instrument*, it is possible that the results of all conceivable measurements are determined.

The second assumption was locality. A measurement made on α [β] does not modify the preset values of the observables $B(b)$, $B(b')$ [$A(a)$, $A(a')$] of β [α]. If one writes

$$A(a) = s, \qquad A(a') = s'$$
$$B(b) = t, \qquad B(b') = t' \tag{73}$$

where s, s', t, t' are all equal to ± 1, locality means that these four parameters, preassigned by the realistic assumption, are not modified at a distance by measurements. Therefore, if $A(a)$ is measured on an α particle, for example, and the value s is found, the preassigned values t and t', associated with the correlated β particle, are in no way modified.

We are obviously dealing with a realistic and *deterministic* approach, since the result of every possible measurement is predetermined by some concrete properties of the measured objects ("hidden variables"). This does not mean, however, that an active role of the apparatus is excluded, but only that the interaction between object and apparatus is driven to a preset outcome ("result of measurement") by the hidden variables of the object.

As a consequence of these assumptions, the set E of N (α, β) pairs splits into 2^4 subsets with well-defined populations in which the outcome of the four possible measurements is predetermined. Let $E(s, s'; t, t')$ be a subset of E with preset values of the four observables (73), and let $n(s, s'; t, t')$ be its population. Naturally

$$\sum n(s, s'; t, t') = N \qquad (74)$$

where $\sum$ denotes a sum over the 2^4 different sets of values of the dichotomic parameters s, s', t, and t'.

By virtue of the locality assumption, the concrete performance of the measurement of $A(a)$, or of $A(a')$, on the α objects of the subset $E_1 \subset E$ does not in any way modify the preset values of $B(b)$ and $B(b')$ in that subset. In other words, there is no action at a distance modifying $B(b)$ and/or $B(b')$ arising from the measurements of $A(a)$ or $A(a')$ (and *vice versa*).

The *a priori* probabilities

$$\rho(s, s'; t, t') = \frac{1}{N} n(s, s'; t, t') \qquad (75)$$

can therefore be used for the calculation of correlations of concretely performed experiments. Therefore,

$$\begin{aligned}
P(a, b) &= \sum \rho(s, s'; t, t')st \\
P(a, b') &= \sum \rho(s, s'; t, t')st' \\
P(a', b) &= \sum \rho(s, s'; t, t')s't \\
P(a', b') &= \sum \rho(s, s'; t, t')s't'
\end{aligned} \qquad (76)$$

where again $\sum$ denotes a sum over the dichotomic variables s, s', t and t'. It is now a simpler matter to show that

$$|P(a, b) - P(a, b')| \leq \sum \rho(s, s'; t, t')|t - t'| \qquad (77)$$

since $|s| = 1$. Similarly, from $|s'| = 1$ it follows that

$$|P(a', b) + P(a', b')| \leq \sum \rho(s, s'; t, t')|t + t'| \qquad (78)$$

By adding Eqs. (77) and (78) and using the equality

$$|t - t'| + |t + t'| = 2$$

which is a consequence of $|t| = |t'| = 1$, Bell's inequality follows, since Eq. (74) is equivalent to

$$\sum \rho(s, s'; t, t') = 1$$

With Wigner's proof, probabilities entered into the EPR paradox for the first time. They were, however, deduced from a deterministic background, much in the same way as was done by Laplace with his formulation of probability calculus.

One can say that in Wigner's approach the role of the hidden variable λ is played by the preset values of $A(a)$, $A(a')$, $B(b)$, $B(b')$. Remembering (73), it becomes immediately clear that $\rho(s, s', t, t')$ is expected in general to vary if one or more of the parameters a, a', b, b' are modified. This is a very important feature of the probabilistic approach based on local realism, which was unfortunately often forgotten in later developments, based on insufficiently general concepts.

REFERENCES

1. A. EINSTEIN, B. PODOLSKY, and N. ROSEN, *Phys. Rev.* **47**, 777–780 (1935).
2. N. BOHR, *Phys. Rev.* **48**, 696–702 (1936).
3. E. SCHRÖDINGER, *Proc. Camb. Phil. Soc.* **31**, 555–563 (1935).
4. W. H. FURRY, *Phys. Rev.* **49**, 393–399 (1936).
5. W. H. FURRY, *Phys. Rev.* **49**, 476 (1936).
6. E. SCHRÖDINGER, *Proc. Camb. Phil. Soc.* **32**, 446–452 (1935).
7. D. FORTUNATO, *Lett. Nuovo Cim.* **15**, 289–290 (1976).
8. D. BOHM, *Quantum Theory* Prentice Hall, Englewood Cliffs, NJ (1951).
9. D. BOHM and Y. AHARONOV, *Phys. Rev.* **108**, 1070–1076 (1957).
10. C. S. WU and I. SHAKNOV, *Phys. Rev.* **77**, 136 (1950).
11. L. KASDAY, in: *Foundations of Quantum Mechanics* (B. d'Espagnat, ed.), Italian Physical Society, Course IL, Academic Press, New York (1971), pp. 195–210.
12. J. M. JAUCH, in: *Foundations of Quantum Mechanics* (B. d'Espagnat, ed.), Italian Physical Society, Course IL, Academic Press, New York (1971), pp. 20–55.
13. H. J. LIPKIN, *Phys. Rev.* **176**, 1715–1718 (1968).
14. H. P. STAPP, *Nuovo Cim. B* **29**, 270–276 (1975).
15. J. S. BELL, in: *Foundations of Quantum Mechanics* (B. d'Espagnat, ed.), Italian Physical Society, Course IL, Academic Press, New York (1971), pp. 171–181.
16. J. S. BELL, *Physics* **1**, 195–200 (1965).
17. J. F. CLAUSER, M. A. HORNE, A. SHIMONY, and R. A. HOLT, *Phys. Rev. Lett.* **23**, 880–884 (1969).
18. E. P. WIGNER, *Am. J. Phys.* **38**, 1005–1009 (1970).

Chapter 2

Bell's Inequality and Its Elementary Background

Now that the history of the EPR paradox has been considered, its conceptual background will be examined. The underlying philosophy is that of local realism, which, in addition to realism and locality, includes time's arrow. From these natural and intuitive elements Bell's inequality, briefly encountered in Chapter 1, is rigorously deduced. Of course it is grossly violated by quantum mechanics, which is therefore inconsistent with the principles represented in the inequality.

Realism can be opposed to complementarity, according to which quantum systems can manifest different, incompatible features—like an undulatory or corpuscular nature—but do not really possess them all at the same time. The aspect revealed depends on the experiment, and no experiment can implement complementary characteristics together.

Complementarity can be more naturally applied to single than to composite systems; to the latter Einstein, Podolsky, and Rosen can apply their reality criterion, allowing the assignment of elements of reality, which are complementary and hence incompatible according to Bohr, to both subsystems. In fact the very possibility of deducing Bell's inequality runs against complementarity.

Certain well-known approaches[1,2] involve "ontological" violations of Heisenberg's uncertainty relations, which express the limits imposed by complementarity.

2.1. LOCAL REALISM

Local realism is characterized by the following three postulates:

1. *There is an objective reality, which exists independently of observation.* The moon is there when nobody is looking. An atom continues to exist

when it is not being observed. This is the spirit of the assumption, whose "letter" is represented by certain precise criteria of reality, such as the EPR reality criterion. These criteria are capable of experimental test, once associated with the following two assumptions.

2. The interaction between two objects can be made arbitrarily small, even negligible, by increasing their separation. This is the assumption of *locality*.

3. Time has a suitably defined "arrow," which means that all physical propagations go from the past to the future and that, hence, the past cannot be modified.

The validity of local realism, thus formulated, is strongly supported by current scientific knowledge. Quantum mechanics, however, is incompatible with it. This was made clear in 1965 with the violation of Bell's inequality, in which the assumptions of Einstein's realism are represented, and which can be compared both with the predictions of quantum mechanics and with experience. Infinitely many other inequalities have also been discovered, which determine limits that cannot be deduced from Bell's.

Local realism coincides more or less with Einstein's position. Einstein was certainly a realist and would have agreed with postulate 1. He neither bothered to mention action at a distance nor the possibility of influencing the past in the 1935 paper, for such issues had yet to become explicit in quantum mechanics. He no doubt would have considered both too absurd to be addressed. When he did appreciate that the possibility of quantum nonlocality could be seriously entertained, he wrote: "But on one supposition we should, in my opinion, absolutely hold fast: the real factual situation of the system S_2 is independent of what is done with the system S_1, which is spatially separated from the former."[3] He could not "seriously believe in [quantum mechanics] because the theory is incompatible with the principle that physics is to represent reality in time and space, without spookish long-distance effects."[4]

2.1.1. The Reality Postulate

The idea that objects exist independently of human beings and their observations is so obvious that it is more or less accepted by everyone, including most theoretical physicists. It is, however, sometimes dismissed as being merely metaphysical and as forming no part of a sensible scientific attitude. It is not metaphysical, however, for when it is combined with locality and the arrow of time it has many empirical consequences, which could, in principle, be falsified by suitable experiments, in particular, those that test Bell-type inequalities with very accurate instruments.

Access to the world is given by the senses. One has an awareness of sense perceptions, so *their* existence cannot be denied. It can, however, be denied that there is anything beyond these sense perceptions and that a real external world produces them. Admittedly the senses give an incomplete and imperfect image of reality. Any correspondence between sense perceptions and elements of that reality cannot be strictly unambiguous. There is no reason why human perception should be able to penetrate reliably the finest details of nature. Different circumstances can be indistinguishable, and similar circumstances can give rise to very different perceptions. The instruments that assist observation, similarly, can introduce distortions and perturbations.

The realist will claim that the intricacies of this relationship between observation and nature *are no reason to abandon a belief in an external reality* altogether, and to claim that things only exist insofar as they are directly observed, or even that their existence is *created* by observations in some sense. The realist can then wonder how to give this view rigor and even empirical substance rather than leaving it vague and metaphysical. In which situations, he can ask, must the presence of an external reality be necessarily recognized? Which "elements" of that reality can be looked for? Which properties of which systems can be considered undeniably real?

As an example, consider the quantum of nuclear forces, the famous π meson or pion. The mass of the charged pion, known with great accuracy, is $m_\pi = 139.5673 \pm 0.0007 \text{ MeV}/c^2$, or approximately $140 \text{ MeV}/c^2$. One can say that "the mass of the pion is equal to $140 \text{ MeV}/c^2$" because measurements always result in that figure. Therefore, it is assumed that all pions have this mass, even those whose mass has not been measured. If it were not possible to assume this, the mass would be literally *created* by the action of the measuring instrument, and pions would possess no real property corresponding to the observed value of the mass. The assumption that measurements also say something about objects that are similar to but distinct from the ones that are directly examined is commonly made in high-energy experiments, and with great success: for instance, verifying the validity of conservation laws (energy, momentum). It is therefore reasonable to assume that there is a reality independent of measurements behind the value of $140 \text{ MeV}/c^2$ for *any* pion's mass.

It is, however, best to avoid strong assertions such as "the mass of the pion really and objectively *has* a value equal to $140 \text{ MeV}/c^2$, even if it is not measured." This would represent an attitude of naive realism and would be like saying that the green color of grass is real. Rather, it is known that observed colors are at least in part appearances and that objective reality is far closer to the wavelength of the light in question than to the corresponding color sensation. Hence, it is more accurate to say that "there is something real that corresponds to the green color of the grass," while it is at least risky, if not downright wrong, to say that "the grass is really green." For similar reasons it is more correct to assert that "there is something real corresponding to the known of value of the mass of the pion." The same applies to just about any property of matter.

These simple considerations form the basis of the EPR reality criterion:[5]

EPR Reality Criterion. If, without in any way disturbing a system, we can predict with certainty (i.e., with probability equal to unity) the value of a physical quantity, then there exists an element of physical reality corresponding to this physical quantity.

The first part of the reality criterion requires that the prediction be made without disturbing the object in question. It is, for instance, possible to predict the color of a car without looking at it and, hence, without disturbing it at all. It is also possible to predict the value of the mass of the next pion crossing a bubble chamber without interacting with it, and so on. In every case in which the prediction can be made in this way, the EPR reality criterion assigns to the object (automobile, pion, . . .) an *element of reality*, i.e., something real that does not necessarily coincide with the observed property but generates it deterministically when a measurement is made. To the predictable green color of grass there corresponds, for instance, the reality of a wavelength that generates in our eyes the sensation of green and, beyond that, the reality of the chemical structure of chlorophyll that absorbs light of all colors but green.

Let us summarize the principal characteristics of the "element of reality" that Einstein *et al.* introduced with their reality criterion:

1. Its existence does not necessarily depend on an act of measurement. This is consistent with the general hypothesis of realism, which logically precedes the formulation of a reality criterion by means of which single pieces of reality can be recognized in specific situations.
2. It is considered the cause of the result exactly predicted by the measurement, *if* measurement is indeed undertaken. Hence realism and causality are closely connected in the EPR reality criterion.
3. It is also viewed as being determined at least partly by the measured object and not entirely by the measuring apparatus, which has been tested and could, in principle, give results that differ from those predicted. If it can be predicted with certainty that such an apparatus will give a predetermined result interacting with an object, the decisive factor that produces the result of measurement can only be in the object.

This third point does not imply that the apparatus does not play the active role in the measurement process one might expect in quantum physics. Indeed, the EPR reality criterion is only applicable to those particular situations in which the result of a measurement can be exactly predicted even before it is performed. These situations arise when the system is in an eigenstate of the observable in question.

Consider, for instance, an electron described by the wave function

$$\psi = \psi_0 e^{i\mathbf{p}\cdot\mathbf{x}/\hbar}$$

which is a monochromatic plane-wave, eigenfunction of the momentum operator. A measurement of momentum performed on an electron described by ψ will necessarily yield the result $\mathbf{p}$; this can be predicted with certainty. Applying the EPR reality criterion, there must exist an element of reality of the electron corresponding to the predicted value $\mathbf{p}$. This does not necessarily mean that the electron *has* a well-defined momentum equal to $\mathbf{p}$. It is only postulated that the exactly predictable result of the measurement of the physical quantity known as momentum, and defined operatively, has a real (but generally unknown) cause. It remains possible, however, that momentum is a largely conventional entity.

For a realist the EPR reality criterion has a solid conceptual basis. It is, however, open to two kinds of criticism, which are related. To begin with, its applicability is rather limited, as it only recognizes as real, in the atomic domain, what quantum mechanics describes with eigenstates of the measured observable. Why can one not introduce elements of reality for all situations described, more generally, using superpositions? Einstein *et al.* were conscious of this limitation and wrote that their criterion, "while far from exhausting all possible ways of recognizing a physical reality, at least provides us with one such way."

The requirement that the physical quantity be predictable *with certainty* represents the other weakness of the EPR reality criterion: when can absolute certainty ever be reached in an actual physical situation? In practice, all kinds of perturbations, such as the sudden arrival of a cosmic ray, combine to eliminate real certainty. In 1935 certain predictability appeared to be a consequence of quantum theory for eigenstates of the measured observable. The situation has, however, been changed by the theorem of Wigner, Araki, and Yanase,[6-8] which demonstrated the necessity of imperfect measurements, even in principle, if the consistency of quantum theory with the known conservation laws (energy, momentum, angular momentum) is extended to measurement processes.

These two difficulties can be dealt with by means of a probabilistic generalization of the reality criterion, in which the *prediction of a probability* is used to deduce the existence of real physical properties of suitable statistical ensembles. Such properties are generalizations of "elements of reality" and will be considered with the probabilistic treatment of the next chapter.

2.1.2. The Locality Postulate

The idea of "locality" is based on the reasonable belief that the intensity of interaction between objects depends inversely on their separation. If a chair collapses, a distant chair remains unaffected; if a ship sinks in Rio, there will be no effect on a ship in Naples; if a star explodes in Andromeda, our sun continues to shine just as before; if a person has an accident in Vienna, a relative in London will not be physically harmed as a consequence; if an electron is observed in a laboratory, another electron 10 or 10^{10} m away acquires no new property; and so forth. Objects may well interact, to some extent, no matter how far apart they are. This is not being denied. There is always the force of gravity, for instance, which acts between any two objects and diminishes with the inverse square of the separation. It tends to zero as that separation tends to infinity, but never vanishes for any finite distance. Only something much simpler and easily acceptable is being claimed, namely that at large distances interactions have very small effects, so small that in practice they can be neglected in normal experiments.

All known interactions tend to zero as the distance between the interacting objects tends to infinity. Gravitational and electromagnetic forces decrease with the inverse square of the distance, while strong and weak forces fall off even more rapidly, in fact exponentially. It was stated that all "known" interactions tend to zero to distinguish them from certain *hypothetical* interactions, such as the ones, whose intensity is assumed to increase with separation, that would bind quarks together inside nuclear matter. This conjecture is not, however, to be taken too seriously yet—if ever—because it has the character of an *ad hoc* hypothesis, made to explain the failure of all attempts to discover these quarks, which remain unobserved. Thus, no known interaction really violates the idea of locality as formulated. Again:

> Given two separated objects A and B, the modifications of A due to anything that may happen to B (collision, interaction with a third object, disintegration, measurement, ...) can be made arbitrarily small, for any measurable physical quantity, by increasing their separation.

This locality postulate is clearly of very general validity, and there appears to be nothing to which it cannot be applied. The formalism of quantum mechanics is about the only thing that could lead one to question it. Even in this case, however, one is strongly tempted to believe in its validity because, again, the intensity of the four fundamental interactions falls off rapidly with distance.

It is often objected, however, that the formalism of quantum mechanics does not represent a reality in ordinary space, since the Schrödinger equation for two particles yields a wave that propagates in six-dimensional space (configuration space). This in fact is the central issue here. Even in classical statistical mechanics configuration space is often used, but there it represents no more than a useful simplification, as it is equally possible to describe the configuration in ordinary three-dimensional space. Such a description is, on the other hand, impossible in quantum mechanics, since nobody has managed to deduce the Schrödinger equation in configuration space from the corresponding equation in ordinary space. It is sometimes believed, therefore, that the abstract spaces used in quantum mechanics have an irreducible meaning and represent the true atomic reality. The notions of spatial proximity and distance may then lose all meaning, or at least have to be revised radically; and those who embrace locality would become naive realists who do not understand the subtleties of the quantum formalism.

But the mathematical apparatus of quantum mechanics, although useful and accurate in its empirical predictions, has no objective existence outside our theoretical schemes. Those who support the contrary are naive Platonists, incapable of understanding the complexity of the relationship between mathematical formalism and objective reality. Mathematics is a human construction and so, therefore, are all the abstract spaces used in quantum mechanics. Ordinary space-time has instead a more fundamental role, if only because it has imposed itself on our species for reasons of survival, long before physics was developed. Whoever does not regard this as being decisive can be asked to provide a solid scientific reason for the abandonment of space-time. This is a request that nobody, at least so far, can satisfy, because all the proposals of an extraspatial physics have their origin in cultural tendencies and philosophical fashions imported from outside the discipline, without an adequate scientific foundation. Otherwise why would people like Planck, Einstein, Schrödinger, and de Broglie have been opposed to the Copenhagen paradigm? And today the discovery of Bell's theorem has, if anything, aggravated the situation.

2.1.3. The Postulate of Time's Arrow

The third principle that features in an unmetaphysical characterization of realism is the impossibility of modifying the past. If a rocket is sent toward the moon, the moon will be reached a few days after the launch. A flash of light directed at the sun will reach it about 8 min after it was produced on Earth. All the effects generated by a propagation in space and time make themselves felt after the beginning of the propagation. Even these assertions can appear to be of obvious validity to the uninitiated; only those accustomed to extravagant fantasies of contemporary physics will not be surprised if they are contested. Naturally time's arrow has yet to be refuted experimentally.

An abandonment of time's arrow has been proposed more than once recently. Clearly this leads to fantastic scenarios. If propagations from the present toward the past are indeed possible, the past can be modified through choices and decisions made in the present. This is clearly commonplace in science fiction, which may have its appeal, but which should not be embraced for reasons of cultural or philosophical fashion, and only, if ever, once solid scientific grounds have made it necessary—that is, if physics is to continue being, as one would hope, a serious activity that advances by conjecture and refutation. Today, instead, the dogmatism that once prevailed has been supplanted by total anarchy, in which it seems that any idea can form the acceptable basis of a scientific development, irrespective of how well founded it may be. This would be less harmful if it were just a matter of provisional acceptance, pending experimental and conceptual clarification. Unfortunately, however, the boundary between the provisional and the definitive has been blurred, and there are fervent supporters of just about anything, including the idea of a propagation toward the past.

One of the arguments employed in favor of propagations toward the past is that antiparticles are to be considered ordinary particles traveling backward in time. Hence a positron, for instance, would be no more than a normal electron going from the future toward the past. This opinion is, however, hardly consistent with the circumstances under which the positron was discovered, as recounted so clearly by Alvarez[9] (Fig. 2.1):

> ... To illustrate the fact that I did not invent these "rules of physics" this morning, let us recall the single essential ingredient in the discovery of the positron. Most physicists would say that the discovery of the positron involved the observation that an electron-like track in a magnetic cloud chamber bent the wrong way. But that would not be correct, since others had previously seen electron-like tracks curving the wrong way in cloud chambers; the effect was always attributed to electrons going in the opposite direction. In fact, Skobletzyn (the first person to build a magnetic cloud chamber) commented on the strange behaviour of electrons—they occasionally scattered through almost exactly 180°! With hindsight, we now recognize that he was seeing pair production, but he believed that the positron was an electron going the other way.
>
> Anderson's great discovery of the positron rested entirely on the fact that he knew which direction his positron was going; he placed a lead plate in his cloud chamber and saw the particle lose energy and "curl up" after it went through the plate. Many observers had seen particles that were *consistent* with the positron hypothesis, but Anderson was the first one to be able to *reject* all other alternatives. That is why we recognize him as the discoverer of the positron.

Hence, a positron is *not* a normal electron traveling toward the past. If it were it would have to gain energy crossing the lead slab. Given the complexity of the

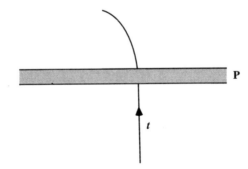

FIGURE 2.1. Anderson's discovery of the positron. The track t is produced in a cloud chamber by a positively charged particle moving upward (not by a negatively charged particle moving downward), because the particle loses energy in the lead plate P, as is clear from the increase of curvature in the magnetic field.

multiple interactions with atoms that take place in the slab, the probability of an energy gain is extremely low, and its regular observation would be entirely inconsistent with all known laws of probability and thermodynamics.

Having dealt with this element of confusion, it can be said that every known phenomenon involves the propagation of matter and/or energy from the past to the future, *and never the other way.* So time's arrow does exist in nature.

2.1.4. Definition of Local Realism

The three postulates of the last sections together represent local realism. They can be summarized as follows:

1. *Reality*: objects and their properties exist independently of human beings and their observations. Suitable reality criteria allow the identification of real physical properties in specific situations.
2. *Locality*: The change in an object due to interactions of any distant object with a third body are very small.
3. *Time's arrow*: All real propagations are directed toward the future. The past therefore cannot be modified.

We have already said that these postulates constitute a form of falsifiable, and hence nonmetaphysical, realism. This is why it is possible to deduce from local realism, with the utmost rigor as shall be seen, the so-called weak inequalities, derived without additional assumptions of any kind. Quantum mechanics predicts marked violations of the weak inequalities for experiments performed with very accurate instruments (for instance, photon detectors whose efficiency is greater than 90%). Such experiments have never been performed, but there is no reason why further technological developments, or perhaps the study of correlated objects other than photons, should not allow an experimental resolution of the EPR paradox in the near future.

Bohr and Heisenberg have suggested that only a *macroscopic* reality can be said to exist independently of observation, and that it makes no sense to speak of any such existence of microscopic objects. Einstein disagreed:[10]

> There is such a thing as the "real state" of a physical system, which exists objectively, independently of any observation or measurement, and which can be described, in principle, with the means of description afforded by physics.... I am not ashamed to make the 'real state of a system' the central concept of my approach.

Locality has been abandoned by many physicists, such as Bohm, Stapp, and Ne'eman, who have invented worlds in which instantaneous actions at a distance between microscopic objects are somehow possible. Time's arrow has been questioned by Wheeler, Costa de Beauregard, and other physicists who have entertained the possibility of an active modification of the past (Chapter 5 is devoted, among other things, to the ideas of these authors and contains the corresponding bibliography).

Common to all these attitudes is the desire to save our current quantum theory, cost what it may. This is understandable, given the great successes of the theory. The real problem, however, is to understand whether the right way out of the paradoxical situation is a denial of very simple and very general assertions that have yet to be refuted empirically or a modification of the existing theory, replacing it with one that preserves its correct empirical predictions while being compatible with Einstein's realism. The real choice between these possibilities can only be provided by experiment. There are two possibilities:

1. The weak inequalities will be found to be in disagreement with experimental results, while quantum mechanics will be confirmed once more.
2. The predictions of quantum mechanics will be violated by these new experiments that do agree with the weak inequalities.

(We shall ignore the pessimistic possibility that experiments violate both the theory and the inequalities!) In the first case the quantum paradigm of modern physics will necessarily have to be accepted; in the second Einstein's realism will have survived a very important test.

2.2. ELEMENTARY PROOF OF BELL'S INEQUALITY

2.2.1. Preliminaries

Again, by "local realism" it is meant that the objects with which the natural sciences deal (galaxies, stars, stones, atoms) somehow exist objectively and independently of human beings and their observations (*realism*) and that they also exist independently of one another (*locality*), in the sense that the farther apart two objects are, the less the physical properties of the one depend on the other. Propagations from future to past must also be excluded (*time's arrow*).

In the next two sections the paradox of Einstein, Podolsky, and Rosen (EPR) will be formulated in the form proposed by Bell: there exists a measurable quantity that assumes two different values according to whether one believes in local realism or in the validity of the empirical predictions of quantum mechanics. The paradox lies in the incompatibility between the two predictions, that is, in the impossibility of interpreting quantum physics from the natural point of view of local realism.

We can begin by considering six points.

1. *Dichotomic physical quantity.* By this is meant a measurable quantity that can assume only two values, $+1$ and -1. In practice every measurable quantity

can be made dichotomic by a suitable redefinition. For instance, a dichotomic weight can be attributed to a person: $+1$ for 70 kg and over, -1 for less; or a dichotomic energy to an electron, which will take on the value $+1$ if its energy is greater than 10 eV and -1 if it is less. This can be done with any quantity.

2. *EPR-type experiment.* In an EPR-type experiment there is a source Σ of pairs of objects (α and β) of any kind and two measuring apparatuses, A_α and A_β, that can measure dichotomic observables on the respective objects. Let us assume that either apparatus can measure one of the four dichotomic physical quantities, a, a', b, b', on the object in question (Fig. 2.2).

Hence in an EPR-type experiment 16 different measurements can be performed: the observable measured by A_α can be chosen in four ways, as can the observable measured by A_β, independently. Assume that either instrument can be set for the chosen observable among the four possible by means of a lever that can assume four positions, each indicating one of the dichotomic quantities (a, a', b, b'). Assume furthermore that the outcome of a single act of measurement performed on an object is displayed on a screen attached to the apparatus. Hence, a $\pm$ on the screen will mean that the outcome of the measurement was ± 1. So every pair of objects α and β, which interact with their respective measuring apparatuses A_α and A_β, will give rise to a pair of signs: $++,+-,-+,--$. In practice, once the measurable quantities have been chosen on the two apparatuses, the measurement will be performed on many successive repetitions of the pairs α and β. Thus, the source Σ will have to be capable of producing a very large number of pairs.

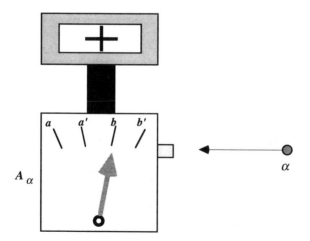

FIGURE 2.2. Experimental apparatus A_α for measuring any of the dichotomic ($=\pm 1$) observables a, a', b, b' on the incoming physical systems α. The result ($+1$) of the last measurement is shown on the upper screen.

3. *Correlation function*. This is the average, over many measurements, of the product of the results obtained by the two apparatuses. $P(x, y)$ will represent the correlation function, where $x = a, a', b, b'$ and $y = a, a', b, b'$ denote the dichotomic physical quantity chosen by the lever on each measuring apparatus (x for A_α and y for A_β). Suppose, for instance, that a and b are measured, and that the pairs of results $(-,+),(+,-),(-,-),(+,-),(-,+),\ldots$ are found, which yield the products $-1, -1, +1, -1, -1,\ldots$. If the frequency of the products $+1$ and -1 over many measurements is the same as in the first five measurements ($+1$ once and -1 four times), the correlation function $P(a, b)$ will be equal to $[1(+1)+4(-1)]/5 = -0.6$. Clearly a correlation function will, by definition, always remain in the interval $(-1, +1)$.

4. *Bell's measurable quantity Δ*. This is a particular linear combination,

$$\Delta = P(a, b) - P(a, b') + P(a', b) + P(a', b') \tag{1}$$

of four correlation functions. Note that only 4 of the 16 correlation functions introduced enter the definition of Δ; furthermore, only the positions a and a' of A_α's lever are taken into consideration, and only b and b' for A_β. The other positions of the levers will, however, be useful in the sequel.

5. *The EPR reality criterion*. Again, a very natural idea that enters the demonstration of Bell's inequality is that physics has to do with *real* objects that exist independently of our observations and instruments. "Real" here is used in the most general and least naive sense possible; if the kind of reality in question is denied, nothing at all remains. Again, rather than leaving matters in generic and intuitive terms, Einstein, Podolsky, and Rosen prefer to rely on a precise statement, which can then be accepted, criticized, or refuted. The following is their reality criterion:

EPR Reality Criterion. If, without in any way disturbing a system, we can predict with certainty (i.e., with probability equal to unity) the value of a physical quantity, then there exists an element of physical reality corresponding to this physical quantity.

Admittedly this is only one of many ways the existence of something real can be recognized or defined, but it derives its great importance from the consequences that can be deduced from it. The reality criterion gets applied to the macroscopic world, sometimes only intuitively, without a precise formulation. For instance, one naturally imagines that a certain "reality" lies behind the possibility of predicting with certainty the result of the next measurement of the length of an object on which many accurate measurements of the same kind were previously performed. In this case, then, there is a real property of the object that manifests itself in the predictable result of the measurement of length. Naive realists will then say that the object is "really" so many centimeters long; others could be aware of anthropomorphic elements not only in the unit of length, which is entirely conventional, but also in the very notion of extension. They will not deny that there is something real behind every measurement whose outcome can be predicted exactly, but they may not wish to commit themselves as to the nature of that "real something." The EPR reality criterion, which will clearly be accepted by the former, is designed also to satisfy the latter as well. Despite the applicability of this criterion to the quantum world, the validity of Bell's inequality has a far more general—in fact universal—validity, if one accepts the natural hypotheses from which it is deduced.

6. *Separability.* Any two objects can be separated enough to make their mutual influences arbitrarily small. To put this otherwise, given any two objects, there exists a distance between them, above which the modification of any physical quantity of either object, owing to their mutual interaction, will be negligibly small. All propagations furthermore go forward in time. Separability, so formulated, can be seen as a consequence of the assumption of locality and of time's arrow discussed in the first section.

This separability hypothesis is based on common sense and on what is now known about interactions. There are four kinds of interaction. The most well known is gravity; there is also an electromagnetic interaction, which binds electrons to nuclei in atoms; a *weak* interaction, responsible for the disintegration of neutrons and other unstable objects; and a *strong* interaction, which binds protons and neutrons together in nuclei.

All interactions fall off rapidly as distance increases. If the quantity "force" is used to measure the interaction, one notices that the gravitational and electric forces diminish like the inverse of the square of the distance, while the weak and strong forces decrease even more rapidly (exponentially). Founded as it is on precise scientific facts and on the existence of an "arrow of time," the assumption of separability would hence appear to be of unquestionable validity.

Bell's inequality is deduced from these six points.

2.2.2. Demonstration of Bell's Inequality

We now discuss a typical EPR-type experiment. It will be useful to make a simplifying assumption to allow an easy application of the reality criterion and the notion of separability, which, however, is unnecessary for the deduction of Bell's inequality, as demonstrations in which it is not made exist as well; one is reported in Section 2.2.4. It will be used to render the argument as clear and elementary as possible. The predictions of quantum mechanics for EPR-type experiments which have been performed *do satisfy the assumption.*

Here, then, is the simplifying assumption:

Assumption. *Every pair of objects α and β has a strong correlation, in the sense that measuring the same physical dichotomic physical quantity on both objects will necessarily yield the same result.*

Suppose the same position, say a (or a', b, or b'), is selected on A_α and A_β. The simplifying assumption of strong correlation implies that repeating measurements on many pairs α and β, pairs of equal results are always obtained: $(-,-)$, $(+,+)$, $(-,-)$, $(-,-)$, $(+,+)$, ... The pair of $+$ signs and the pair of $-$ signs alternate haphazardly with frequencies that depend on the nature of the chosen dichotomic quantity. The product of the two results, however, is always $+1$. Hence, the correlation functions $P(a,a)$, $P(a',a')$, $P(b,b)$, $P(b',b')$ are all exactly equal to $+1$.

Finally, we demonstrate Bell's inequality. Again, it is reached by applying the reality criterion of Einstein, Podolsky, and Rosen and the idea of the separability of the two objects of every pair. Suppose it has been checked repeatedly that the emitted pairs really are strongly correlated. Faith in the regularity of nature ensures that the same will be true of the next pair. Given, then, this strong correlation, it is enough to look at the screen of A_α (see Fig. 2.3) and read the sign on it to be able to predict with certainty that the same sign is on A_β's screen. In many cases strong correlation does not depend on time; that is, it holds even if A_α and A_β are not equidistant from Σ. The prediction can be referred to the case in which the distance between Σ and A_α is less than the distance between Σ and A_β: if a sign appears on A's screen, the same sign will appear later on the other screen. These are exactly the conditions required for the application of the EPR reality criterion, so it can be concluded that *there exists an element of physical reality* of β that corresponds to the value, predictable with certainty, of the dichotomic physical quantity to be measured on the object.

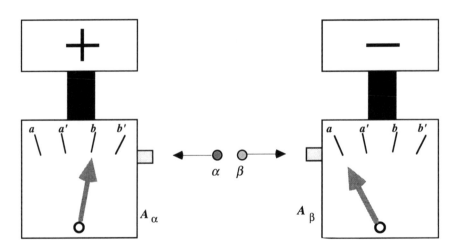

FIGURE 2.3. Experimental apparatuses A_α and A_β measuring one of the dichotomic observables a, a', b, b' on the incoming physical systems α and β, respectively. The results ($+1$ and -1) of the last measurements are shown on the upper screens.

If, for instance, $P(b, b)$ is measured, it can be concluded that there exists an element of reality of β, which is the *cause* of the result of the future measurement of b on β. Let $R_+(b)$ be the element of reality that predetermines the result $b = +1$, and let $R_-(b)$ be the element that determines $b = -1$.

The element of reality is, in any case, an objectively real property of β that has to be viewed as existing even if the future measurement of b on β is not performed. Were this not the case, having accepted the EPR reality criterion and the idea of separability, one would have to conclude that $R_+(b)$ is created retroactively by the future measurement performed on β, which is implausible and excluded by local realism. So $R_+(b)$ belongs to β even if the measurement of b is not performed, provided an ample foundation of previous observations has made it clear that the process in question involves a strong correlation which does not depend on time.

In the same conditions it can be concluded that $R_+(b)$ belongs to β *even if the measurement of b is not performed on* α. Otherwise, if $R_+(b)$ can be attributed to β only when a measurement of b is performed on α, the measurement on α would create $R_+(b)$ at a distance for β. This conflicts with separability. Therefore, *for every pair of objects* α *and* β *emitted by* Σ *there exists an element of reality,* $R_+(b)$ *or* $R_-(b)$, *which predetermines the result,* $+1$ *or* -1, *respectively, of a possible future measurement of b performed on* β.

This conclusion holds for the objects β of all the pairs emitted by the source, even if no measurement of b is performed on α or on β. Similar arguments can be applied to the other correlation functions, $P(b', b')$, $P(a, a)$, $P(a', a')$. Furthermore the whole argument is symmetric in α and β. So the elements of reality $R_+(a)$, $R_+(a')$, $R_+(b)$, $R_+(b')$ can be attributed to both α and β. Some, however, are superfluous, which means the argument can be simplified. Only $R_+(a)$ and $R_+(a')$ will be attributed to α and $R_+(b)$ and $R_+(b')$ to β. This means there are four different signs to choose, so there are 16 different ways of making the choice. Therefore there exist 16 different kinds of pairs (α, β), described in Table 2.1.

In the last column the numerical value of the measurable quantity Δ for every one of the 16 types of pairs is given. A single example is provided here, and the other 15 cases are left to the reader.

For the pairs of type 1 the elements of reality are $R_+(a)$, $R_+(a')$, $R_+(b)$, and $R_+(b')$. This means that the quantities a and a' of α and the quantities b and b' of β are all predicted to have the value $+1$ for these pairs. These values are *locally* determined by objective properties of the corresponding objects on which the dichotomic quantities are measured; that is, they do not change if the measured quantity of the *other* object changes. Hence the correlation function $P(a, b)$ is predicted to have the value $+1$ for type 1 pairs. For the same pairs—and the same reason—the correlation functions $P(a, b')$, $P(a', b)$, $P(a', b')$ are also predicted to have the value $+1$. Combining the correlation functions as in (1), one obtains $\Delta = +2$, just as in Table 2.1. The other values of Δ displayed in Table 2.1 are obtained similarly.

TABLE 2.1. Elements of Reality for the 16 Types of EPR Pairs

	Elements of reality for α		Elements of reality for β		Δ
1	$R_+(a)$	$R_+(a')$	$R_+(b)$	$R_+(b')$	$+2$
2	$R_+(a)$	$R_+(a')$	$R_+(b)$	$R_-(b')$	$+2$
3	$R_+(a)$	$R_+(a')$	$R_-(b)$	$R_+(b')$	-2
4	$R_+(a)$	$R_-(a')$	$R_+(b)$	$R_+(b')$	-2
5	$R_-(a)$	$R_+(a')$	$R_+(b)$	$R_+(b')$	$+2$
6	$R_+(a)$	$R_+(a')$	$R_-(b)$	$R_-(b')$	-2
7	$R_+(a)$	$R_-(a')$	$R_+(b)$	$R_-(b')$	$+2$
8	$R_-(a)$	$R_+(a')$	$R_+(b)$	$R_-(b')$	-2
9	$R_+(a)$	$R_-(a')$	$R_-(b)$	$R_+(b')$	-2
10	$R_-(a)$	$R_+(a')$	$R_-(b)$	$R_+(b')$	$+2$
11	$R_-(a)$	$R_-(a')$	$R_+(b)$	$R_+(b')$	-2
12	$R_+(a)$	$R_-(a')$	$R_-(b)$	$R_-(b')$	$+2$
13	$R_-(a)$	$R_+(a')$	$R_-(b)$	$R_-(b')$	-2
14	$R_-(a)$	$R_-(a')$	$R_+(b)$	$R_-(b')$	-2
15	$R_-(a)$	$R_-(a')$	$R_-(b)$	$R_+(b')$	$+2$
16	$R_-(a)$	$R_-(a')$	$R_-(b)$	$R_-(b')$	$+2$

In general, the pairs emitted by Σ have to be considered a disordered mixture of the 16 possible types with *a priori* unknown frequencies. But it is not necessary to know the frequencies with which the various kinds of pairs are produced by Σ, because the table indicates that Δ measured on the ensemble of all the pairs emitted will be a weighted average of eight $+2$ values and as many -2 values. As such, Δ will never leave the $(+2, -2)$ interval, which implies that $|\Delta| \leq 2$, which is Bell's inequality.[11]

This last result is absolutely general and can be applied to pairs of galaxies, pairs of stars, pairs of atomic objects, . . . , once the dichotomic physical quantities have been defined. The interest of Bell's inequality lies, among other things, in this universality of the result. One could wonder whether its application to atomic objects is impossible on account of the incompatibility of the measurement of quantum observables described by noncommuting linear Hermitian operators. The incompatibility of the observables means, however, that they *cannot be measured simultaneously*, and nothing implies that this prevents their simultaneous *predetermination*, even though only one can be measured at a time on a given quantum object.

2.2.3. Incompatibility with Quantum Mechanics

In the last two sections it was never necessary to specify the nature of the two objects α and β: The extreme generality of the assumptions allows the deduction of Bell's inequality for arbitrarily complicated objects, of any size and

FIGURE 2.4. Two quantum systems α and β used in the experimental study of the EPR paradox can be represented as wave packets localized in different regions of space and having (group) velocities directed in opposite directions. Spherical waves have been adopted in the figure.

nature. *No phenomenon in any of the natural sciences is known to violate Bell's inequality.* The only exception is represented by the predictions of quantum mechanics for EPR-type experiments performed with atomic and subatomic systems.

In the atomic and subatomic world there exist numerous examples of correlated pairs of the kind considered in Fig. 1.1. Every object (molecule, atom, particle) capable of decaying into two new objects α and β represents an example:

The nitrogen–oxygen molecule (NO) can decay spontaneously from an excited state to a state in which the two free atoms (N and O) propagate in opposite directions.

The π^0 meson can disintegrate into two gamma rays.

Certain excited atomic states generate the simultaneous emission of two optical photons; and so on.

The extraordinary fact that has given such importance to Bell's inequality is that there exist situations in which quantum mechanics predicts unambiguously that the quantity Δ must assume the value $\Delta = 2\sqrt{2} = 2.828$, a violation of Bell's inequality ($\Delta \leq 2$) by more than 40%. Hence, the existing quantum theory is incompatible with the natural philosophy of local realism, and this holds at an empirical level. This is the essence of the EPR paradox, which can be resolved, in favor of quantum mechanics or local realism, by appropriate experiments. Contrary to some opinions, experiments performed so far have not decided the matter. This important point will be discussed in detail in Chapter 3.

It is interesting to observe that the violations of local realism contained in quantum mechanics have nothing to do with difficulties related to the representation of measurable quantities ("observables") by Hermitian operators. Indeed the operators of two different quantum objects commute anyway, and hence no influence on β generated by a measurement performed on α is implied (and *vice versa*).

2.2.4. Original Proof of Bell's Inequality

It is useful to consider again the simplest formal proof of Bell's inequality, essentially as given by Bell in his 1970 Varenna lecture,[12] which was sketched in Section 1.8. It is assumed that in an EPR experiment dichotomic observables $A(a) = \pm 1$ and $B(b) = \pm 1$ are measured on the two particles α and β, respectively, moving in opposite directions, as in Fig. 2.4. These observables depend on instrumental parameters a and b (polarizers' axes, directions of magnetic fields, etc.) that can be varied. In practice, only two observables $[A(a)$ and $A(a')]$ are of interest for the α particles, and two $[B(b)$ and $B(b')]$ for the β particles. In general, it is expected that $A(a)$ and $A(a')$ are incompatible and hence cannot be measured at the same time, and that the same holds for $B(b)$ and $B(b')$.

It is assumed that hidden variables belonging to α and β fix the outcome of all possible measurements. These hidden variables are collectively represented by λ, assumed to vary in a set Λ with a probability density $\rho(\lambda)$. The normalization condition

$$\int_\Lambda d\lambda\, \rho(\lambda) = 1 \qquad (2)$$

holds.

Thus one can write

$$A(a, \lambda) = \pm 1; \quad A(a', \lambda) = \pm 1; \quad B(b, \lambda) = \pm 1; \quad B(b', \lambda) = \pm 1$$

meaning that, given λ, every one of the four dichotomic observables assumes a well-defined value. The correlation function $P(a, b)$ is defined, as usual, as the average product of two dichotomic observables. In the hidden-variable approach,

$$P(a, b) = \int_\Lambda d\lambda\, \rho(\lambda) A(a, \lambda) B(b, \lambda)$$

One can obviously write

$$|P(a, b) - P(a, b') + P(a', b) + P(a'b')|$$

$$\leq \int_\Lambda d\lambda\, \rho(\lambda)\{|A(a, \lambda)|\, |B(b, \lambda) - B(b', \lambda)| + |A(a', \lambda)|\, |B(b, \lambda) + B(b', \lambda)|\}$$

$$= \int_\Lambda d\lambda\, \rho(\lambda)\{|B(b, \lambda) - B(b', \lambda)| + |B(b, \lambda) + B(b', \lambda)|\}$$

since $|A(a, \lambda)| = |A(a', \lambda)| = 1$. But the moduli of $B(b, \lambda)$ and $B(b', \lambda)$ are also equal to 1, so

$$|B(b, \lambda) - B(b', \lambda)| + |B(b, \lambda) + B(b', \lambda)| = 2 \qquad (3)$$

From (2) and (3) it follows that

$$|P(a, b) - P(a, b') + P(a', b) + P(a', b')| \leq 2$$

which, again, is Bell's inequality.

This proof is based on a general form of realism because the hidden variable λ is thought to belong objectively to the real physical systems α and β. It is also based on locality for three reasons: (1) the dichotomic observable $A(a, \lambda)$ $[B(b, \lambda)]$ of particle α [β] does not depend on the parameter b [a] of the other experimental apparatus; (2) the probability density $\rho(\lambda)$ does not depend on a, b; and (3) the set Λ of possible λ values does not depend on a, b. The time arrow assumption is also implicit in the independence of $\rho(\lambda)$ on a, b. In principle, the choices of the values of the instrumental parameters could be made when the particles α and β are in flight from the source to the analyzers. Without a time arrow one could nevertheless conceive that "really" the particles propagate backward in time from the detector to the source and that they modify the source in ways depending on a and b.

The hidden variable λ is essentially the same as the EPR "element of reality," even though its existence is postulated without invoking the EPR reality criterion. Therefore, this proof does not require a strong correlation of the dichotomic observables, as would, for example, be expressed by the condition $A(a, \lambda) = -B(a, \lambda)$, valid for all λ. Such a strong correlation was needed in the proof of Bell's inequality in Section 2.2.2 because there the reality criterion was central to the proof of the inequality.

2.3. OTHER FORMULATIONS OF THE EPR PARADOX

Bell's inequality is the single most important expression of the incompatibility between local realism and quantum mechanics, but it is not the only one. Here two other expressions will be considered; in Chapter 3 other inequalities deduced from the principles of local realism, but not reducible to Bell's, will be examined as well.

2.3.1. Systems without a State

Given two subsystems α and β, if the state of α is described by $|\psi\rangle \in \mathcal{H}^\alpha$ and that of β by $|\varphi\rangle \in \mathcal{H}^\beta$, the state of the whole system (α, β) will be described by the product $|\psi\rangle|\varphi\rangle$, which belongs to the space $\mathcal{H}^\alpha \otimes \mathcal{H}^\beta$ arising from the multiplication of $\mathcal{H}^\alpha$ and $\mathcal{H}^\beta$. The linearity of this product space allows state vectors to be added to produce others. Consider, for instance, the vector

$$|\eta\rangle = \frac{1}{\sqrt{2}}\{|\psi_1\rangle|\varphi_2\rangle + |\psi_2\rangle|\varphi_1\rangle\} \tag{4}$$

where $|\psi_i\rangle \in \mathcal{H}^\alpha$ and $|\varphi_j\rangle \in \mathcal{H}^\beta$ ($i, j = 1, 2$). In the absence of superselection rules (which apply only seldom and in well-known circumstances) the vector $|\eta\rangle$ will describe a possible state of the pair (α, β). Indeed as $|\psi_1\rangle|\varphi_2\rangle$ and $|\psi_2\rangle|\varphi_1\rangle$ describe possible states of the pair, so will any linear combination of them, particularly (4).

The impossibility of writing (4) as a product represents a first expression of the EPR paradox. Any vector $|\psi\rangle|\varphi\rangle$ can be expanded with respect to the sets $|\psi_i\rangle$ and $|\varphi_j\rangle$:

$$|\psi\rangle = \sum_\mu a_\mu |\psi_\mu\rangle, \quad |\varphi\rangle = \sum_\nu b_\nu |\varphi_\nu\rangle, \qquad \mu, \nu = 1, 2, 3, \ldots$$

whence, as tensor multiplication is linear in either factor,

$$|\psi\rangle|\varphi\rangle = \sum_{\mu\nu} a_\mu b_\nu |\psi_\mu\rangle|\varphi_\nu\rangle \qquad (5)$$

To try to reproduce (4) one has to assume that indices other than 1 and 2 do not figure in (5); in other words, only a_1, a_2, b_1, b_2 do not vanish. Then (5) becomes

$$|\psi\rangle|\varphi\rangle = a_1 b_1 |\psi_1\rangle|\varphi_1\rangle + a_1 b_2 |\psi_1\rangle|\varphi_2\rangle + a_2 b_1 |\psi_2\rangle|\varphi_1\rangle + a_2 b_2 |\psi_2\rangle|\varphi_2\rangle$$

As the term $|\psi_1\rangle|\varphi_1\rangle$ does not arise in (4), the product $a_1\, b_1$ must vanish, which means that $a_1 = 0$ or $b_1 = 0$. But in the first case $|\psi_1\rangle$ disappears from (5), in the second $|\varphi_1\rangle$. Hence no choice of coefficients allows the vector (5) to be reduced to (4). But (5) is the most general vector that can be expressed as a product of a vector describing α and another β. Therefore *state vector (4) cannot be factorized.*[13]

This is a serious physical problem, not a mere technicality. In quantum mechanics the only relationship between the theoretical formalism and objective reality is given by the state vector (or wave function). Even if this description is very peculiar, most people would agree that *some* reality lies behind the formalism. Therefore the idea that there is something real corresponding to the "quantum object" presupposes that some vector $|\varphi\rangle$ should describe it. But if either system has a state vector, the two taken together should be described by a product. *So, in a sense, if α is real and β is real, the state vector describing both must be a product.* Hence if the vector in the tensor product space is not a product of vectors in the factor spaces, an independent reality cannot be attributed to the two subsystems. The formalism seems to indicate that the pair (α, β) is real, but neither α nor β is separately. This is a serious problem because in quantum mechanics one sometimes has to use nonfactorizable state vectors to describe subsystems very far apart from one another and, hence, presumably in no position to interact.

2.3.2. Determination at a Distance of the State

Suppose the pair (α, β) is made up of two spin-$\frac{1}{2}$ particles in the singlet state; in other words, it is described by the vector

$$|\eta_0\rangle = \frac{1}{\sqrt{2}}\{|u_+\rangle|v_-\rangle - |u_-\rangle|v_+\rangle\} \tag{6}$$

where

$$\sigma_3|u_\pm\rangle = \pm|u_\pm\rangle \qquad \text{and} \qquad \tau_3|v_\pm\rangle = \pm|v_\pm\rangle$$

σ_3 (τ_3) being the third Pauli matrix for particle α (β).

An important feature of the singlet state is the invariance under rotations, which will now be illustrated. To begin with, the most general spin component for particle α is represented by

$$\boldsymbol{\sigma} \cdot \hat{\mathbf{n}} = \begin{pmatrix} n_3 & n_1 - in_2 \\ n_1 + in_2 & -n_3 \end{pmatrix} \tag{7}$$

which is easily shown to have eigenvalues ± 1 if $|\hat{\mathbf{n}}| = 1$. The corresponding (normalized) eigenvectors will have the general form

$$|u_+^{(n)}\rangle = c_1|u_+\rangle + c_2|u_-\rangle \qquad \text{and} \qquad |u_-^{(n)}\rangle = \tilde{c}_1|u_+\rangle + \tilde{c}_2|u_-\rangle \tag{8}$$

where $|c_1|^2 + |c_2|^2 = |\tilde{c}_1|^2 + |\tilde{c}_2|^2 = 1$. The values of the four coefficients in (8) can be determined by solving the eigenvalue equations of the matrix (7). One gets

$$c_1 = \frac{1 + n_3}{\sqrt{2(1 + n_3)}}, \qquad c_2 = \frac{n_1 + in_2}{\sqrt{2(1 + n_3)}} \tag{9}$$

and

$$\tilde{c}_1 = \frac{1 - n_3}{\sqrt{2(1 - n_3)}}, \qquad \tilde{c}_2 = \frac{-n_1 - in_2}{\sqrt{2(1 - n_3)}} \tag{10}$$

Equations (10) and (9) differ, one will see, only as to the sign of $\hat{\mathbf{n}}$. This makes sense, since the equation with eigenvalue -1 of $\boldsymbol{\sigma} \cdot \hat{\mathbf{n}}$ is the same as the equation with eigenvalue $+1$ of $\boldsymbol{\sigma} \cdot (-\mathbf{n})$.

The corresponding spin matrix for particle β is

$$\boldsymbol{\tau} \cdot \hat{\mathbf{n}} = \begin{pmatrix} n_3 & n_1 - in_2 \\ n_1 + in_2 & -n_3 \end{pmatrix} \tag{11}$$

Here again the eigenvalues are ± 1 and the corresponding eigenvectors are

$$|v_+^{(n)}\rangle = c_1|v_+\rangle + c_2|v_-\rangle \qquad \text{and} \qquad |v_-^{(n)}\rangle = \tilde{c}_1|v_+\rangle + \tilde{c}_2|v_-\rangle \tag{12}$$

with the same coefficients (9) and (10). Inverting (8) and (12) one obtains

$$|u_+\rangle = c_1|u_+^{(n)}\rangle + \tilde{c}_1|u_-^{(n)}\rangle, \qquad |u_-\rangle = c_2|u_+^{(n)}\rangle + \tilde{c}_2|u_-^{(n)}\rangle \qquad (13)$$

and

$$|v_+\rangle = c_1|v_+^{(n)}\rangle + \tilde{c}_1|v_-^{(n)}\rangle, \qquad |v_-\rangle = c_2|v_+^{(n)}\rangle + \tilde{c}_2|v_-^{(n)}\rangle \qquad (14)$$

respectively. Notice that the normalization is correct because from (9) and (10) it follows that $|c_1|^2 + |\tilde{c}_1|^2 = |c_2|^2 + |\tilde{c}_2|^2 = 1$.

Substituting these expressions into the singlet, one obtains

$$
\begin{aligned}
|\eta_0\rangle &= \frac{1}{\sqrt{2}}\left\{\left[c_1|u_+^{(n)}\rangle + \tilde{c}_1|u_-^{(n)}\rangle\right]\left[c_2|v_+^{(n)}\rangle + \tilde{c}_2|v_-^{(n)}\rangle\right]\right. \\
&\quad \left. -\left[c_2|u_+^{(n)}\rangle + \tilde{c}_2|u_-^{(n)}\rangle\right]\left[c_1|v_+^{(n)}\rangle + \tilde{c}_1|v_-^{(n)}\rangle\right]\right\} \\
&= \frac{1}{\sqrt{2}}\left\{[c_1\tilde{c}_2 - c_2\tilde{c}_1]|u_+^{(n)}\rangle|v_-^{(n)}\rangle + [\tilde{c}_1 c_2 - \tilde{c}_2 c_1]|u_-^{(n)}\rangle|v_+^{(n)}\rangle\right\} \\
&= \frac{1}{\sqrt{2}}\left\{|u_+^{(n)}\rangle|v_-^{(n)}\rangle - |u_-^{(n)}\rangle|v_+^{(n)}\rangle\right\} \qquad (15)
\end{aligned}
$$

This expansion is structurally identical to (6), but contains the eigenstates of the spin component along direction $\hat{\mathbf{n}}$. From (8) and (9) one has

$$c_1\tilde{c}_2 - c_2\tilde{c}_1 = -(\tilde{c}_1 c_2 - \tilde{c}_2 c_1) = \frac{n_1 + in_2}{\sqrt{1 - n_3^2}}$$

which is a phase factor and can be neglected. So the *singlet state is rotationally invariant*. This has several important consequences. To begin with, two observations can be made.

(1) The spin observables written out in full are

$$S_n = \frac{\hbar}{2}\boldsymbol{\sigma}\cdot\hat{\mathbf{n}}, \qquad T_n = \frac{\hbar}{2}\boldsymbol{\tau}\cdot\hat{\mathbf{n}} \qquad (16)$$

It can be deduced from (15) that if a measurement of S_n gives the result $\pm\hbar/2$, the measurement of T_n on particle β of the same pair will give the opposite result $\mp\hbar/2$. It is important to note that this prediction always applies; time does not come into it at all. The measurements do not have to be made at the same time. The evolution of the observables in question is

$$S_n(t) = e^{iH^\alpha t}S_n(0)e^{-iH^\alpha t}$$

where H^α is the Hamiltonian of the first particle. Since, at least in nonrelativistic quantum mechanics, $[S_n(0), H^\alpha] = 0$ for a free particle and $e^{iH^\alpha t}$ is a function of H^α, we have

$$S_n(t) = S_n(0)e^{iH^\alpha t}e^{-iH^\alpha t} = S_n(0)$$

Of course a similar relation holds for the other particle. So the anticorrelation of spin measurements follows from the fact that the singlet belongs to the null space of the operator $[S_n(t) + T_n(t')]$ for all t, t'; in other words

$$[S_n(t) + T_n(t')]|\eta_0\rangle = 0 \tag{17}$$

This means that if a measurement of S_n at time t produces the result ± 1, a measurement of T_n on particle β of the same pair at any other previous or future time t' must give ∓ 1.

In relativistic physics the matter is not so straightforward because the spin of a Dirac particle does not commute with the Hamiltonian, but the result is the same.

(2) Two *different* operators like (7) do not commute. Indeed it follows from (7) itself that

$$[\boldsymbol{\sigma} \cdot \hat{\mathbf{n}}, \boldsymbol{\sigma} \cdot \hat{\mathbf{n}}'] = 2i\boldsymbol{\sigma} \cdot \hat{\mathbf{n}} \times \hat{\mathbf{n}}'$$

which vanishes only is $\hat{\mathbf{n}}$ and $\hat{\mathbf{n}}'$ are parallel or antiparallel. This means that it is impossible to measure the two corresponding observables (which are the spin components of the particle along directions $\hat{\mathbf{n}}$ and $\hat{\mathbf{n}}'$) at the same time. According to the Copenhagen interpretation they cannot both be well defined in the state of the system. In other words, $\boldsymbol{\sigma} \cdot \hat{\mathbf{n}}$ and $\boldsymbol{\sigma} \cdot \hat{\mathbf{n}}'$ cannot have predetermined values if $\hat{\mathbf{n}} \neq \pm\hat{\mathbf{n}}'$ for any state vector describing particle α.

Since the singlet is in the null space of $(\boldsymbol{\sigma} + \boldsymbol{\tau}) \cdot \hat{\mathbf{n}}$ for all $\hat{\mathbf{n}}$, the paradox can be reformulated in the following terms. Take a statistical ensemble of (α, β) pairs in the singlet state, whose α-particles always travel leftward and whose β-particles travel rightward. The apparatuses can be set up very far from the source to prevent the spatial parts of the wave functions from overlapping. If observable T_n is measured on the β-particles, it is easy to demonstrate that the average value $\langle\eta_0|T_n|\eta_0\rangle = 0$. So a random sequence of $+\hbar/2$ and $-\hbar/2$ with equal frequency will result from measurements. If, for a given β-particle, one finds $+\hbar/2$ $\{-\hbar/2\}$, the result is, according to the axioms of quantum mechanics, a well-defined value. So the state of particle β becomes $|v_+^{(n)}\rangle$ $\{|v_-^{(n)}\rangle\}$. Taking account of the anti-correlation predicted by (17) the state of the pair must be

$$|u_-^{(n)}\rangle|v_+^{(n)}\rangle \quad \{|u_+^{(n)}\rangle|v_-^{(n)}\rangle\} \tag{18}$$

Performing measurements on the β-particles, the $\hat{\mathbf{n}}$-component of the spin of the α-particles gets determined at a distance. Had another $\hat{\mathbf{n}}$ been chosen, the reality of another component would have been manifested. Therefore the reality of the α's is determined at a distance. It can be shown that this has no statistical consequences because a mixture of the two vectors (18) in equal proportions has a density matrix that does not depend on *n*. It is, nevertheless, a disturbing result for a realistically minded physicist.[14]

2.4. CRITICISM OF THE COPENHAGEN APPROACH

Bohr's reply to the argument of Einstein, Podolsky, and Rosen is founded on his notion of complementarity; this has been looked at in Section 1.2. Complementarity is, in turn, closely related to Heisenberg's uncertainty relations. But, useful as they are in a description of measurements and their outcomes, it is misleading to ascribe *ontological* significance to these notions. The drift from the empirical to the ontological was dangerously easy, and indeed encouraged, in the atmosphere of positivistic zeal that prevailed during the early development of quantum theory.

The fact that only one of two complementary features can be *manifested* at a time does not, as we shall soon see, mean that both cannot be *possessed* together; likewise the perturbation of a particle's position by the measurement of its momentum does not mean that the unobserved particle cannot describe a trajectory. But if one wishes to attribute reality only to the measured, and dismiss all else as being metaphysical, or just not there, then a particle's position—not what we know about it, *the position itself*—is lost when momentum is measured. The position is not perturbed, or changed, but destroyed. It vanishes and becomes undefined, indeterminate.

2.4.1. Heisenberg's indeterminacy relations

Every physics student learns *Gedankenexperimente*, qualitative considerations and mathematical proofs that establish the validity of Heisenberg's indeterminacy relations:

$$\Delta x \Delta p_x \geq h, \quad \Delta y \Delta p_y \geq h, \quad \Delta z \Delta p_z \geq h \tag{19}$$

It is only seldom pointed out, however, how arbitrary Heisenberg's arguments are. The conclusion that one cannot simultaneously *measure* the position and momentum of an electron with arbitrarily small errors does not imply that such quantities have no physical meaning and must be considered objectively undetermined. Only a positivist attitude, prevalent in Heisenberg's times, can lead to such an inference. Moreover one can show that the validity of the uncertainty relations still allows one to calculate simultaneous values of position and momentum *in the past* with any desired accuracy. Heisenberg wrote in 1930:[15]

Then for these past times $\Delta x \Delta p$ is smaller than the usual limiting value, but this knowledge of the past is of a purely speculative character, since it can never (because of the unknown change in momentum caused by the position measurement) be used as an initial condition in any calculation of the future progress of the electron and thus cannot be subjected to experimental verification. It is a matter of personal belief whether such a calculation concerning the past history of the electron can be ascribed any physical reality or not.

Heisenberg's own "personal belief," which has since become the dominant point of view, was that one should always refrain from attributing physical reality to any retrodicted coordinate or momentum values. Fortunately it is possible to disagree. Indeed it is possible to calculate position and momentum in the past with a precision better than Eq. (19) would seem *a priori* to allow. The simplest way to understand this point is to repeat Heisenberg's argument accompanying the above quotation in his book. Consider an electron initially described by the plane wave

$$\psi(x, y, z) = \psi_0 e^{i\mathbf{p}\cdot\mathbf{r}/\hbar} \tag{20}$$

for which $\Delta p_x = \Delta p_y = \Delta p_z = 0$: the electron's momentum $\mathbf{p}$ is known exactly. At time t_0 let a position measurement be made on the electron with finite errors Δx, Δy, and Δz, that is, with the results

$$x = x_0 \pm \Delta x/2, \quad y = y_0 \pm \Delta y/2, \quad z = z_0 \pm \Delta z/2 \tag{21}$$

Position and momentum certainly satisfy (19) at any time $t \geq t_0$, because momentum suffers a perturbation and becomes unknown when position is measured. Thinking of the situation existing before t_0, one can instead draw a very different conclusion. Given (20), at all times $t \leq t_0$ the electron had a perfectly determined momentum $\mathbf{p}$, implying

$$\Delta p_x = \Delta p_y = \Delta p_z = 0 \tag{22}$$

Clearly, for $t \leq t_0$ also the velocity $\mathbf{v} = \mathbf{p}/m$ was exactly determined and known. Therefore the electron position can be retrodicted from (21), valid at $t = t_0$, to have been for all $t \leq t_0$:

$$x(t) = x_0 \pm \frac{\Delta x}{2} - v_x(t_0 - t)$$

$$y(t) = y_0 \pm \frac{\Delta y}{2} - v_y(t_0 - t) \tag{23}$$

$$z(t) = z_0 \pm \frac{\Delta z}{2} - v_z(t_0 - t)$$

The errors of the retrodicted position are the same Δx, Δy, and Δz that applied at time t_0. But finite errors for position and zero errors for momentum, as in (22), imply

$$\Delta x \Delta p_x = \Delta y \Delta p_y = \Delta z \Delta p_z = 0 \tag{24}$$

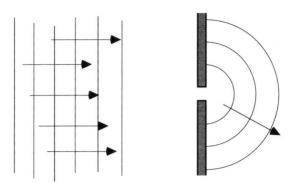

FIGURE 2.5. Heisenberg's *Gedankenexperiment* on the single-slit diffraction of electrons is normally used to establish the validity of the indeterminacy relations. As shown in the text, with a different approach it can also be used to violate the indeterminacy relations in the past.

in violation of (19). This is the essence of Heisenberg's argument leading to a violation of the indeterminacy relations in the past (Fig. 2.5).

There is only one way to refute (24) and it consists of Heisenberg's positivistic assumption—present enough in the previous quotation—that it is forbidden to attribute position and momentum to an unobserved electron, such as the electron for which the results (22) and (23) were calculated at time $t \leq t_0$. This is a philosophy with a Machian flavor, which is not even fashionable today, after the errors made by Mach, in his obstinate refusal to accept the reality of atoms, have been amply recognized.

Today it is concretely possible to adopt a realistic attitude and to claim that (24) shows that Heisenberg's indeterminacy relations are violated in the objective reality of the past and then, by natural extension, at all times. After all the future does not yet exist, the present is no more than a boundary, and the past is the only reality we can claim to know with some accuracy. It is therefore also concluded that the plane wave (20), which does not contain any information about the electronic particle localization, describes only an average reality of the electron's motion and that Heisenberg's relations (19) must be interpreted as scatter relations valid in a statistical ensemble of similarly prepared electrons. The individual electron, instead, has a well-defined reality to which Eq. (24) applies, and quantum mechanics is incomplete in the Einstein–Podolsky–Rosen sense.

Ideas of this type have been adopted by some contemporary authors.[16,17]. Wesley[16] gives several examples in which Heisenberg's relations (19) appear to be grossly violated (a photon in a living cell, a photon received by a pocket radio, the electron in the hydrogen atom, a photon in the tip of a scanning light microscope, the electron in beta decay, . . .). To give the flavor of these arguments consider the beta decay of a tritium nucleus of radius 1.7×10^{-13} cm. At the instant the electron leaves the nucleus it is localized, say, in a sphere two times larger than the nucleus. Then

$$\Delta x \leq 3.4 \times 10^{-13} \text{ cm}$$

The kinetic energy of the emitted electron is about $T = 18.6$ keV, which corresponds to a momentum of $p = \sqrt{2mT} \cong 7.4 \times 10^{-18}$ g-cm/s. The uncertainty in the momentum Δp of the electron as it leaves the tritium nucleus is then certainly less than p itself; therefore,

$$\Delta p / \hbar \leq 7.0 \times 10^9 \text{ cm}^{-1}$$

The simultaneous uncertainties in position and momentum then yield

$$\Delta x \Delta p / \hbar \leq 2.4 \times 10^{-3} \ll 1$$

In considering the foregoing derivation, one should not forget the strictly realistic attitude that lies behind it. Wesley's electron is a particle that must be viewed as being localized in space at every instant of time: it is not clear whether it should also be imagined as endowed with a well-defined momentum. In all cases it is, of course, no wonder that such an entity can be shown to violate (19). Actually one knows that an electron has a dual nature (wave + particle), but it seems legitimate to ask a question concerning the particle alone.

Interesting are also the arguments presented by Croca,[17] who believes that the path to challenge Heisenberg's relations is now open for two reasons: One is mathematics with the development of wavelet analysis, more general than the usual Fourier analysis; the other is a purely technological breakthrough in the techniques of microscopic imaging, allowing experimental resolutions that greatly violate the usual maximum resolution limit

$$\Delta x \cong \frac{\lambda}{2 \sin \varepsilon} \tag{25}$$

This new generation of microscopes is typified by the scanning tunneling microscope, for which Binnig and Rohrer[18] received the Nobel prize. Recent developments allowed position to be measured with a resolution $\approx \lambda/50$, in sharp violation of (25).

In Croca's mathematical argument it is implied that all quantum particles are contained in a wave having a Gaussian amplitude centered at point x_0 where the particle is concretely located:

$$\psi_k(x) \cong A_0 \exp[-(x - x_0)^2 / 2\sigma_0^2] \exp[ikx] \tag{26}$$

The essential point is that (26) is the most elementary wave to be considered and that the fundamental relation $p = hk$ between particle momentum p and "wave number" k is assumed to hold only with the k appearing in (26), not with the different wave numbers arising from a Fourier analysis of (26). This means that the elementary wavelet (24) describes a particle having $\Delta x = \sigma_0$ and $\Delta p = 0$, thus a particle violating Heisenberg's relation from the start. Also some well-known quantum postulates are here violated, but the argument is interesting in showing that Heisenberg's relations are valid if one accepts quantum mechanics: a future modification of the theory could well lead to an overcoming of Heisenberg's limit. This is just the opposite of what is usually believed concerning the relationship between the "uncertainty principle" and quantum theory, about which Feynman wrote:[19]

> The uncertainty principle "protects" quantum mechanics. Heisenberg recognized that if it were possible to measure the momentum and the position simultaneously with a greater accuracy, the quantum mechanics would collapse. So he proposed that it must be impossible. Then people sat down and tried to figure out ways of doing it, and nobody could figure out a way to measure the position of anything—a screen, an electron, a billiard ball, anything—with any greater accuracy. Quantum mechanics maintains its perilous but accurate existence.

In the light of the previous considerations it appears that Feynman's statement is justified only if one sticks to a strict positivistic philosophy ("the only reality consists of acts of measurement"). For a realist the completeness of quantum theory has already collapsed at least since 1935.

2.4.2. Complementarity for Space-Time and Causality

To begin with, Bohr's formulation of complementarity for space-time and causality will be considered. There is an instinctive tendency, rooted in direct macroscopic experience, to describe events in space and time *and* establish causal relations between them; these have become fundamental "categories" of understanding and communication.

It is natural to describe events as taking place in three-dimensional space and evolving in time. So one says that a seminar takes place in the main hall of the physics building starting at 10 A.M. tomorrow, that a picture was taken in June at 10,000 m above Iceland, that a gift was bought in India during the last winter vacation, and so on. The use of space-time in classical physics is common practice and represents an extrapolation of daily life conceptions to the new domains considered by modern science. The scientific description is quantitative and as unambiguous as possible: particular sets of space-time coordinates are introduced to this end.

A similar extension from ordinary life to physics is used for the idea of causality, again with a more precise and quantitative formulation in the scientific case; an evolution is causal if it takes place according to well-defined rules. For Bohr the most important of these rules are the conservation laws of energy and momentum since they identify a small part of the set of all conceivable processes by allowing the actual *reality* of only those in which the conserved quantities of an isolated system remain unchanged. Neglecting for simplicity other conservation laws (angular momentum, electric charge, . . .), one can regard a *causal* description as being one according to the conservation of energy and momentum, at least in classical physics.

The physicist who studies the phenomena of the quantum domain will naturally *try* to use his or her macroscopic preconceptions and will *try* to describe atomic processes as taking place in space-time and according to the rules of energy–momentum conservation. However, he or she will soon find that it is not possible to do so, because two quantum observables are often described by noncommuting operators, which implies not only that they cannot be measured simultaneously but also that the measurement of one of them in general destroys all previous knowledge concerning the other.

The question of space localization (position measurement) and causality implementation (momentum measurement) represents a good illustration of complementarity. Precise space localization of a quantum system can be obtained by measuring position with infinite precision ($\Delta x = 0$). Immediately after such a measurement the wave function becomes the δ-function $\delta(x - x_0)$, if x_0 is the obtained value of position, and the particle can be thought to be really *at* x_0. But a δ-function can be written as a superposition of the whole range of plane waves, each with the same weight, which means that nothing is known about momentum p, or, in other words, that $\Delta p = \infty$. One therefore loses all the knowledge about p available before position was measured, so a definite momentum should not be attributed to the particle; here the momentum literally does not exist. But this also means there is no evidence about the conservation of momentum, e.g., in collision processes. Bohr concludes that a spatial localization leads to the abandonment of causal description.

Symmetrically, with a different experiment, one can choose a causal description instead. This can be done by measuring momentum p with infinite precision ($\Delta p = 0$), and it is well known that momentum is found to be conserved whenever checked. The wave function of the quantum system becomes a plane wave after such a measurement, and one can *now* describe a quantum system as having a well-defined momentum. However, as the squared modulus of a plane wave is constant, nothing at all is known about position, since the same probability density is assigned to all of space ($\Delta x = \infty$). In this way the localization in space is completely lost, and one cannot insist on the reality of a well-defined position. Bohr concludes that a concrete implementation of causal description forces the physicist to abandon the description of the quantum system as localized in space.

The two possibilities *that form the essence of classical physics* (space localization *and* causal description) are thus seen in the quantum domain to be mutually incompatible, to exclude one another. Bohr concludes that in quantum physics it is, in principle, impossible to represent atomic processes as developing causally in space and time and that such an impossibility can be attributed to the finite value of the quantum of action h (Planck's constant). It is in fact the nonzero value of h that generates the noncommutative algebra of quantum operators and which makes a simultaneous measurement of two of them impossible. Bohr considered each one of the two incompatible descriptions as *necessary* for physics: understanding and culture are formed in the macroscopic world where space-time and causality are absolutely necessary. Therefore, one cannot avoid thinking that atomic events take place in space-time and develop causally, but the impossibility of using the two representations together must be accepted. In Hegelian terms it can be said that the thesis (space-time) and the antithesis (causality) can never be united in a synthesis but must always remain opposed and irreconcilable. The latter observation is essential for an understanding of the historical and cultural roots of complementarity; this problem will, however, be discussed in Section 2.4.6.

2.4.3. Complementarity for Spin Components

Bohr's impossibility argument against a simultaneous description in realistic terms applies to any two incompatible observables. Consider, without loss of generality, the case of a spin-$\frac{1}{2}$ particle. It is well known that any two of the spin-component operators S_1, S_2, S_3 do not commute. This means that their corresponding observables S_1, S_2, S_3 cannot be measured at the same time. Let the spin-$\frac{1}{2}$ particle in question be in the initial quantum state

$$|\psi_i\rangle = \begin{pmatrix} 1 \\ 0 \end{pmatrix} \qquad (27)$$

for which S_3 is well defined and equal to $+\hbar/2$. The state (27) can be written as a superposition

$$\begin{pmatrix} 1 \\ 0 \end{pmatrix} = \frac{1}{\sqrt{2}}\left\{\frac{1}{\sqrt{2}}\begin{pmatrix} 1 \\ 1 \end{pmatrix}\right\} + \frac{1}{\sqrt{2}}\left\{\frac{1}{\sqrt{2}}\begin{pmatrix} 1 \\ -1 \end{pmatrix}\right\} \qquad (28)$$

of the two (normalized) eigenspinors of the S_1 operator. According to the quantum-mechanical rules a measurement of the S_1 component can give two results

$$S_1 = \pm\frac{\hbar}{2} \qquad (29)$$

and the final spin state (after measurement) must correspondingly become

$$|\psi_f\rangle = \text{either} \quad \frac{1}{\sqrt{2}}\begin{pmatrix} 1 \\ 1 \end{pmatrix} \quad \text{or} \quad \frac{1}{\sqrt{2}}\begin{pmatrix} 1 \\ -1 \end{pmatrix} \qquad (30)$$

where the spinors on the right side are eigenstates of the S_1 operator. Each of them can in turn be written as a superposition of the two eigenstates of the S_3 operator. Consider, for instance, the equality

$$\frac{1}{\sqrt{2}}\begin{pmatrix} 1 \\ 1 \end{pmatrix} = \frac{1}{\sqrt{2}}\begin{pmatrix} 1 \\ 0 \end{pmatrix} + \frac{1}{\sqrt{2}}\begin{pmatrix} 0 \\ 1 \end{pmatrix} \tag{31}$$

where the normalized spinors on the right side are the eigenstates of the S_3 operator. Since the moduli of the coefficients in superposition (31) are equal, the two possible S_3 eigenstates are equiprobable. The S_3 component must therefore be considered completely unknown in state (31). The same applies to the second spinor on the right side of (30). The complementarity principle leads to the conclusion that the implementation (by measurement) of the reality of S_1 leaves S_3 completely undetermined. The argument is symmetrical in both values (1 and 3) of the index; S_3 can become known if measured, but in such a case it is S_1 that becomes necessarily completely undetermined. One can then conclude, with Bohr, that S_1 and S_3 are complementary aspects of reality: in some cases S_1 can be considered real, in other cases S_3, but never both in the same physical situation. By "real" one means "predetermined by objective properties of the observed system."

In the following sections it will be shown that the previous conclusions are by no means inevitable or "necessary" because complementarity only *postulates* rather arbitrarily certain difficulties that, in fact, do not exist, since a very natural description of physical reality can be given in which all different spin components are considered real at the same time.

2.4.4. General Quantum Predictions for Spin-$\frac{1}{2}$

An arbitrary spin observable of a spin-$\frac{1}{2}$ particle is represented in quantum mechanics by the most general 2×2 Hermitian matrix Σ, which can be written

$$\Sigma = \alpha I + \boldsymbol{\beta} \cdot \boldsymbol{\sigma} \tag{32}$$

where α, β_1, β_2, β_3 are four real constants, I is the 2×2 unit matrix, and σ_1, σ_2, σ_3 are the usual 2×2 Pauli matrices, related to the spin-component operators S_i in the usual way:

$$S_i = \frac{\hbar}{2}\sigma_i \qquad (i = 1, 2, 3) \tag{33}$$

The most general spin state is

$$|\psi\rangle = \begin{pmatrix} c_1 \\ c_2 \end{pmatrix} \tag{34}$$

for which the normalization condition

$$\langle\psi|\psi\rangle = |c_1|^2 + |c_2|^2 = 1 \tag{35}$$

is assumed satisfied. It is easy to show that to every spinor of this last kind one can associate a direction in space specified by a unit vector $\hat{\mathbf{n}}$, in the sense that the eigenvalue equation

$$\hat{\boldsymbol{\sigma}} \cdot \hat{\mathbf{n}} \begin{pmatrix} c_1 \\ c_2 \end{pmatrix} = \begin{pmatrix} c_1 \\ c_2 \end{pmatrix} \tag{36}$$

can be satisfied for a suitable $\hat{\mathbf{n}}$ with the eigenvalue $+1$, as shown. This is the inverse of the problem dealt with in Section 2.3.2. The components of $\hat{\mathbf{n}}$ can be written in terms of c_1, c_2, and *vice versa*. Apart from an irrelevant phase factor one has, as in (9),

$$c_1 = \frac{1 + n_3}{\sqrt{2(1 + n_3)}}, \qquad c_2 = \frac{n_1 + in_2}{\sqrt{2(1 + n_3)}} \tag{37}$$

By changing the sign of $\hat{\mathbf{n}}$ everywhere in (36) and (37), one obtains

$$\boldsymbol{\sigma} \cdot \hat{\mathbf{n}} \begin{pmatrix} \tilde{c}_1 \\ \tilde{c}_2 \end{pmatrix} = -\begin{pmatrix} \tilde{c}_1 \\ \tilde{c}_2 \end{pmatrix} \tag{38}$$

where

$$\tilde{c}_1 = \frac{1 - n_3}{\sqrt{2(1 - n_3)}}, \qquad \tilde{c}_2 = \frac{-n_1 - in_2}{\sqrt{2(1 - n_3)}} \tag{39}$$

as in (10). The two eigenvalues (± 1) of $\boldsymbol{\sigma} \cdot \hat{\mathbf{n}}$ and the corresponding eigenvectors have therefore been obtained. These results can be extended without any difficulty to the operator (32), since one can set

$$\hat{\mathbf{n}} = \frac{\boldsymbol{\beta}}{|\boldsymbol{\beta}|} \tag{40}$$

and note that Eqs. (36) and (38) become eigenvalue equations of the operator $\boldsymbol{\sigma} \cdot \boldsymbol{\beta}$ with eigenvalues $\pm|\boldsymbol{\beta}|$, respectively. The extra term containing α in (32) is proportional to the unit matrix and can only contribute an additional term to the eigenvalues. Therefore,

$$\Sigma \begin{pmatrix} c_1 \\ c_2 \end{pmatrix} = (\alpha + |\boldsymbol{\beta}|) \begin{pmatrix} c_1 \\ c_2 \end{pmatrix}, \qquad \Sigma \begin{pmatrix} \tilde{c}_1 \\ \tilde{c}_2 \end{pmatrix} = (\alpha - |\boldsymbol{\beta}|) \begin{pmatrix} \tilde{c}_1 \\ \tilde{c}_2 \end{pmatrix} \tag{41}$$

Hence, the eigenvalues of the operator Σ are

$$\Sigma_{1,2} = \alpha \pm |\boldsymbol{\beta}| \tag{42}$$

and the corresponding eigenvectors have already been found.

It will next be supposed that the spin-$\frac{1}{2}$ systems in question are initially in the (normalized) state

$$|\psi_0\rangle = \begin{pmatrix} c_{01} \\ c_{02} \end{pmatrix} \tag{43}$$

for which an observable $\Sigma_0 = \alpha_0 I + \boldsymbol{\beta}_0 \cdot \boldsymbol{\sigma}$ can obviously be found having $|\psi_0\rangle$ as eigenvector with eigenvalue $+1$, namely such that

$$c_{01} = \frac{1 + n_{03}}{\sqrt{2(1 + n_{03})}}, \qquad c_{02} = \frac{n_{01} + i n_{02}}{\sqrt{2(1 + n_{03})}} \tag{44}$$

if $\mathbf{n}_0 = \boldsymbol{\beta}_0 / |\boldsymbol{\beta}_0|$. By using the standard expressions for the Pauli matrices the expectation value of operator (32) in state (43) can easily be calculated:

$$\langle \psi_0 | \Sigma | \psi_0 \rangle = \alpha + \beta_3 (|c_{01}|^2 - |c_{02}|^2) + \mathrm{Re}\, \{ c_{01}^* c_{02} (\beta_1 - \beta_2) \} \tag{45}$$

If account is taken of (44), this last result becomes

$$\langle \psi_0 | \Sigma | \psi_0 \rangle = \alpha + \hat{\mathbf{n}}_0 \cdot \boldsymbol{\beta} = \alpha + \frac{\boldsymbol{\beta}_0 \cdot \boldsymbol{\beta}}{|\boldsymbol{\beta}_0|} \tag{46}$$

One must also have

$$\langle \psi_0 | \Sigma | \psi_0 \rangle = p_1 \Sigma_1 + p_2 \Sigma_2 \tag{47}$$

if $p_{1,2}$ are the *a priori* probabilities in the state $|\psi_0\rangle$ of the eigenvalues $\Sigma_{1,2}$, respectively, of course, with

$$p_1 + p_2 = 1 \tag{48}$$

Therefore it follows, by comparing (46) and (47), that

$$p_{1,2} = \frac{1}{2} \left[1 \pm \frac{\boldsymbol{\beta}_0 \cdot \boldsymbol{\beta}}{|\boldsymbol{\beta}_0| |\boldsymbol{\beta}|} \right] \tag{49}$$

In the next section it will be shown that realistic and causal models exist that reproduce exactly the quantum-mechanical predictions for $\Sigma_{1,2}$ and for $p_{1,2}$.

Before coming to this matter, there is a final quantum-mechanical prediction which is useful to recall. Considering the operator defined by Eq. (32) and a second similar operator

$$\Sigma' = \alpha' I + \boldsymbol{\beta}' \cdot \boldsymbol{\sigma} \tag{50}$$

it is a simple matter to calculate the commutator

$$[\Sigma, \Sigma'] = 2i\boldsymbol{\sigma} \cdot \boldsymbol{\beta} \times \boldsymbol{\beta}' \tag{51}$$

Obviously this will vanish only if the vectors $\boldsymbol{\beta}$ and $\boldsymbol{\beta}'$ are parallel (or antiparallel). It can be concluded that two spin observables are always incompatible if they refer to directions of space forming angles different from 0 and from π.

2.4.5. Simultaneous Reality of Different Spin Components

It will now be shown that a model of the physical reality of spin measurements exists, that does not contradict the empirical predictions of quantum theory in any way, with the following features:

(i) A causal description of quantum spin measurements is provided that is in complete agreement with the predictions of standard quantum theory for the "results" $\Sigma_{1,2}$ and for their probabilities $p_{1,2}$.

(ii) All spin observables are simultaneously predetermined by objective properties λ of the measured systems.

(iii) There exist physical properties α, β_1, β_2, β_3 of the measuring apparatus that are essential in defining logically the measured observable and in building up concretely the result of the measurement.

(iv) During the measurement the λ's undergo a statistical redistribution. When properly chosen this redistribution allows one to reproduce exactly the quantum-mechanical predictions for subsequent spin measurements.

(v) It is physically impossible to measure simultaneously two observables that in quantum mechanics are described by noncommuting operators.

The model assumes that quantum-mechanical state vectors describe some concrete physical situations, and not one's knowledge of the atomic world, as Pauli and Heisenberg preferred to believe. The model can be based on the following ideas. First it is assumed that the physical system is a spinning sphere with angular momentum λ forming an angle θ_0 with respect to the direction n_0 specifying the quantum state (43) as in (44). Second, an apparatus built for measuring the observable represented in quantum mechanics by the matrix (32) is assumed to work as follows:

1. It measures the sign μ of the projection of λ on β by "squashing" λ onto β and registering the value $+1$ (-1) if the two vectors after squashing are found to be parallel (antiparallel) [after which it modifies λ according to rules to be specified later].

2. It multiplies the result so obtained by β.

3. It adds α to the new result.

The outcome of these three operations is

$$\alpha + |\beta| \cdot \mu \tag{52}$$

where

$$\mu = \text{Sign}\{\boldsymbol{\lambda} \cdot \boldsymbol{\beta}\} \tag{53}$$

Obviously (52) will in all cases coincide with one of the two quantum-mechanical eigenvalues $\Sigma_{1,2}$ given in (42).

In order to reproduce the quantum-mechanical probabilities as well it is necessary to define within this model the "state" represented in quantum mechanics by the spinor (43)–(44). In a realistic approach the quantum state is represented in much more concrete terms than allowed by a purely abstract representation such as (43). In fact this spinor can be assumed to describe a statistical ensemble of spinning spheres with $\boldsymbol{\lambda}$ vectors distributed according to a density function independent of the azimuthal angle ϕ_0 and given in terms of the polar angle θ_0 by

$$\rho(\theta_0) = \begin{cases} \dfrac{1}{\pi} \cos\theta_0 & \text{if } 0 \leq \theta_0 \leq \dfrac{\pi}{2} \\ 0 & \text{otherwise} \end{cases} \tag{54}$$

which is never negative and satisfies

$$\int_0^{2\pi} d\phi_0 \int_0^{\pi} d\theta_0 \, \sin\theta_0 \, \rho(\theta_0) = 1 \tag{55}$$

It is only necessary to show that the negative content of the complementarity principle can be overcome, and the previous model is perfectly adequate for this purpose, as will immediately be shown. It is not necessary to believe that the model is a perfect representation of reality, since there must be other realistic models which lead to the same empirical consequences as the one here considered. The choice of the best model for spin-$\frac{1}{2}$ goes beyond the purposes of this chapter.

Given the rather mechanical nature of steps 2 and 3 of the previously defined measurement process, the probability of finding the result $\Sigma_1 = \alpha + |\boldsymbol{\beta}|$ is obviously equal to the probability of finding $\boldsymbol{\lambda} \cdot \boldsymbol{\beta}$ positive. (See Fig. 2.6.) The outcome of the assumed "squashing process" of $\boldsymbol{\lambda}$ on $\boldsymbol{\beta}$ is deterministically fixed by the initial sign of the scalar product of these two vectors. Considering a system of (polar, azimuthal) angles, one can assume the directions of $\boldsymbol{\beta}_0$ and $\boldsymbol{\beta}$ to be specified by $(0, 0)$ and $(\theta_1, 0)$, respectively, and that of $\boldsymbol{\lambda}$ by the running angles (θ_0, ϕ_0). Note that θ_1 is the angle between $\boldsymbol{\beta}_0$ and $\boldsymbol{\beta}$. The condition for obtaining $\boldsymbol{\lambda} \cdot \boldsymbol{\beta}$ positive is then

$$\cos\theta = \cos\theta_0 \cos\theta_1 + \sin\theta_0 \sin\theta_1 \cos\phi_0 \tag{56}$$

From the latter condition one obtains

$$\cos^2\theta_0 = \frac{\tan^2\theta_1 \cos^2\phi_0}{1 + \tan^2\theta_1 \cos^2\phi_0} \tag{57}$$

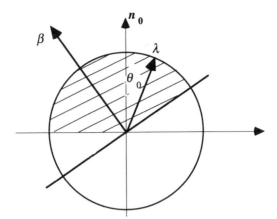

FIGURE 2.6. Hidden-variable model for spin-$\frac{1}{2}$. The "eigenvalue" $+\hbar/2$ of the $\hat{n}_0$ spin component is represented by a statistical distribution of λ's in the upper half-plane, with density (54). If a given λ of a set with such a distribution falls in the dashed area, the measurement of the spin component $\sigma \cdot \beta$ will also give $+\hbar/2$.

From (54) and (55) it follows that

$$p_1 = \int_0^{2\pi} d\phi_0 \int_0^1 d\cos\theta_0 \cdot \frac{1}{\pi} \cos\theta_0 \cdot \Theta[\cos\theta_0 \cos\theta_1 + \sin\theta_0 \sin\theta_1 \cos\phi_0]$$

(58)

where Θ is the Heaviside step function:

$$\Theta[x] = \begin{cases} 1 & \text{if } x \geq 0 \\ 0 & \text{if } x < 0 \end{cases}$$

If the right side of (57) is called $L(\theta_1, \phi_0)$, and (56) is taken into account, it is not difficult to obtain

$$p_1 = \frac{1}{2} + \frac{1}{2\pi} \int_{\pi/2}^{3\pi/2} d\phi_0 \, [\cos^2\theta_0]_{L(\theta_1,\phi_0)}^1$$

(59)

which is easily transformed into

$$p_1 = \frac{1}{2} + \frac{1}{2\pi} \int_{\pi/2}^{3\pi/2} d\phi_0 \, \frac{1}{1 + \tan^2\theta_1 \cos^2\phi_0} = \frac{1}{2}[1 + \cos\theta_1]$$

(60)

Given the definition of θ_1, this coincides with the first of Eqs. (49). Since $p_1 + p_2 = 1$, even the second result (49) is necessarily obtained from a correctly normalized model like this.

The model therefore reproduces quantum-mechanical eigenvalues and their probabilities for all possible measurements and for the most general state, as desired. There is, however, a further point that needs to be investigated, that of subsequent spin measurements on the same particle. Quantum mechanics states that a "reduction of the state vector" takes place during the measurement, as a consequence of which the statistical properties of the considered set of particles are modified. The same result can be obtained by assuming that during measurement the angular momentum vector $\boldsymbol{\lambda}$ undergoes a change of direction. In the cases in which the value Σ_1 of the measured observable has been obtained this change of direction is assumed to take place in two steps: (i) As has been seen, $\boldsymbol{\lambda}$ is squashed onto the direction $\boldsymbol{\beta}$ contained in the apparatus and defining the measured observable. (ii) Immediately after $\boldsymbol{\lambda}$ is redistributed in space around the direction $\boldsymbol{\beta}$ with a probability independent of the azimuthal angle and dependent only on the (polar) angle θ between $\boldsymbol{\lambda}$ and $\boldsymbol{\beta}$ introduced in (56) and given by

$$\rho_1(\theta) = \begin{cases} \dfrac{1}{\pi} \cos\theta & \text{if } 0 \le \theta \le \dfrac{\pi}{2} \\ 0 & \text{otherwise} \end{cases} \tag{61}$$

Where the result Σ_2 is obtained a similar redistribution of $\boldsymbol{\lambda}$ takes place, but around the direction $-\boldsymbol{\beta}$, so the new density of vectors $\boldsymbol{\lambda}$ becomes

$$\rho_2(\theta) = \begin{cases} \dfrac{1}{\pi} |\cos\theta| & \text{if } \dfrac{\pi}{2} < \theta \le \pi \\ 0 & \text{otherwise} \end{cases} \tag{62}$$

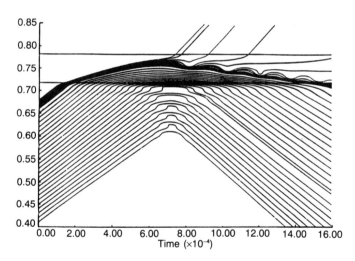

FIGURE 2.7. Tunnel effect reinterpreted. A potential barrier is represented by the two horizontal lines. In the de Broglie–Bohm model particles are surrounded by an objectively existing wave. The figure represents a statistical ensemble of particles, all embedded in the same wave packet, but with different positions. Every individual particle describes a trajectory. As one can see, particles in front (in the rear) of the wave packet cross (are reflected by) the barrier. Time on abscissa, space (only one dimension) on the ordinate. Calculations made by Dewdney and Hiley.[34]

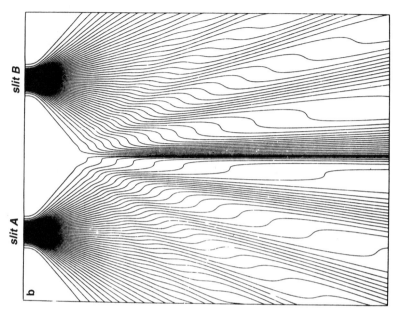

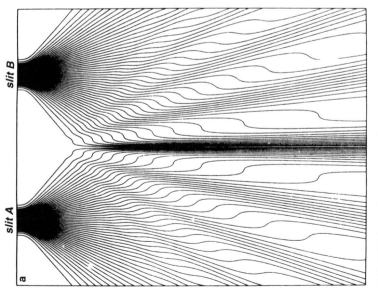

FIGURE 2.8. Particle trajectories for the two-slit experiment. The first screen (with slits) is in the upper part of the drawings, the second screen in the lower. The figures represent a statistical ensemble of particles, all embedded in the same wave packet but crossing the slits in different positions. (a) Particle trajectories for the two-slit arrangement using the de Broglie–Bohm quantum potential approach; (b) particle trajectories for the Aharonov–Bohm effect with the shift arising from the presence of an enclosed magnetic field. Calculations made by Philippidis *et al.*[(35)]

In this way the statistical set of spin-$\frac{1}{2}$ particles during measurement undergoes precisely those statistical modifications that ensure the validity of the usual quantum-mechanical predictions also for subsequent spin measurements. It is physically impossible to squash the $\boldsymbol{\lambda}$ vectors simultaneously onto two non-collinear directions: therefore in this model, as in quantum mechanics, two observables referring to two noncollinear directions in space are incompatible and cannot be measured simultaneously.

At this point it becomes apparent that this realistic model for spin-$\frac{1}{2}$ reproduces exactly all the quantum-mechanical predictions while violating the complementarity principle. Obviously Bohr's formulation of quantum theory was not demanded by strong scientific evidence, if the principle of complementarity can be rejected without rejecting quantum theory itself. The arbitrariness of Bohr's reasoning turns out to be in the *assumption* that if two observables cannot be measured simultaneously then they cannot be predetermined by objectively existing physical properties of the measured system. It is all as simple as that!

The conference *New Theories in Physics* was held in Warsaw in 1938. In the proceedings of this meeting one can see that von Neumann[20] exposed his famous theorem against "hidden variables" and that Bohr[21] commented on it by expressing his highly appreciative opinion and his admiration. He also pointed out that one of his papers, in more elementary ways, had arrived at essentially the same conclusion. The paper Bohr had in mind was clearly the one he had written three years before[22] as an answer to Einstein, Podolsky, and Rosen.[5] This illuminating connection with von Neumann's theorem shows that Bohr believed that his formulation of complementarity could provide an "impossibility proof" against Einstein's causal space-time program. It should therefore not be surprising that arguments not very different from those used against von Neumann-type theorems[23] can be used against complementarity as well.

2.4.6. Historical Roots

In order to understand Bohr's complementarity one has to start from the observation made at the end of Section 2.4.2: Bohr can by no means be considered a Hegelian. It is true that he accepts the presence of contradictory elements in the physical theory, e.g., those of particle and of wave for the description of the same physical system. The particle is localized within a very small region of space and its presence requires the postulation of a vacuum for separating different particles and therefore introduces a discontinuity. The wave occupies instead a large region of space—the whole universe if it is a plane wave—and its presence allows for a continuous description of reality. Since no object can at the same time be large and small, says Bohr, particle and wave are two contradictory elements that are clearly related to another pair of opposites, those of continuity and discontinuity. But this partial acceptance of Hegelian views does not at all imply that for Bohr the contradictions can be resolved in higher syntheses: on the contrary, the avoidance of such resolutions can be considered one of the central tendencies of his life! One of his pupils, Leon Rosenfeld, described the situation in these terms:[24]

> While the great masters (Planck, Einstein, Born, and Schrödinger) were vainly trying to eliminate the contradictions in Aristotelian fashion by reducing one aspect to another, Bohr realized the futility of such attempts. He knew that we have to live with this dilemma... and that the real problem was to refine the language of physics so as to provide room for the coexistence of the two conceptions.

Living with the dilemma is clearly something very different from trying to solve it. If one understands that the Danish philosophers Søren Kierkegaard (1813–1855) and Harald Høffding (1843–1931) entertained exactly the same opinion about the impossibility of resolving contradictions, the nature of Bohr's complementarity becomes clear: complementarity was imported into physics from a secondary branch of philosophy, Danish existentialism! A few quotations can give the flavor of such a connection. In *Die Zerstörung der Vernunft*, György Lukáks devoted 60 pages to Kierkegaard and wrote, for example:[25]

> Kierkegaard fights against Hegel by breaking the living dialectical unity of the two contradictory elements and bringing them into a complete isolation, and by swelling them into independent metaphysical principles.

Now this is exactly what Bohr does with the ideas of particle and of wave and with his refusal to accept the Einstein–de Broglie picture in which the two aspects coexist objectively and constitute the observed physical system.[26] It is extremely difficult to imagine that such a strange idea (that the contradictions regarding physical reality cannot be resolved as a matter of principle!) could be conceived several times independently; hence Jammer writes:[27]

> There can be no doubt that the Danish precursor of modern existentialism and neo-orthodox theology, Søren Kierkegaard, through his influence on Bohr, affected also the course of modern physics to some extent.

One can add that the influence of Kierkegaard on Bohr was not limited to some marginal features of quantum theory but concerned the very central idea of complementarity. In fact, for Kierkegaard contradictions in life and in natural sciences are rigid and impossible to overcome. This conviction went together with sharp attacks against Hegel's rationalism and his idea that contradictions can always be synthesized (i.e., resolved) at a higher level. On the contrary, thought Kierkegaard, contradictions are irreconcilable and constitute alternatives that mutually and definitely exclude one another. Thus, e.g., in a person's life the conflict between, on the one hand, the desires and the needs of the subject (the *thesis*) and, on the other, the concrete situation in which that person is constrained to move (the *antithesis*) provokes a state of permanent anguish that can never be resolved, in Hegelian fashion, in a synthesis of a different quality. A similar situation is encountered in the natural sciences, even though it is fair to add that Kierkegaard was much more interested in theological, spiritual, and psychological problems than in scientific ones. He termed his philosophy "qualitative dialectic."

As seen in Section 2.4.2, Bohr thought that an element of irrationality is introduced into physics by the finite value of the quantum of action h, so scientists must always oscillate between space-time and causality rather than being able to use both of them. Similarly he thought it impossible to use the conceptions of particle and of wave at the same time, but that one must again oscillate between the former and the latter in order to "explain" different experimental facts. Also for Bohr, one is in the presence of contradictory opposites that are mutually irreconcilable and thus not utilizable in a synthesis that could provide a complete description of the essential aspects of quantum reality.

If the conceptual continuity between Kierkegaard and Bohr is clear, it is difficult to find a *direct* connection between them, partly because the former died half a century before the latter began university. There is, however, plenty of evidence of important indirect connections, the key figure being Høffding. Among the early reconstructions of Bohr's cultural background, one should read a little known but important paper by Nils Svartholm[28] that provides a reconstruction of the relationships between Kierkegaard and Høffding and between Høffding and Bohr. More recently there have been studies by Favrholdt,[29], Faye,[30] and Moreira.[31]

The importance of Høffding for his own cultural formation has been so described by Bohr[32] in a posthumous homage (1931) during a talk at the Royal Academy of Sciences and Letters:

> [I had] the privilege of having been in close contact with Høffding since my early youth, because my father was an intimate friend of him and I had, at all stages of my life, the possibility of benefiting from his true scientific and philosophical spirit....

Favrholdt adopted a radical internalist vision of the history of science and tried to describe complementarity as an *inevitable* consequence of the experimental developments and of the quantum-mechanical formalism. This thesis has been proven wrong here since it has been shown that quantum-mechanical predictions can be reproduced by a realistic model while violating the complementarity principle. Therefore complementarity was not inevitable. The present conclusion is more in line with the results of Faye and Moreira, who point out that Høffding postulated the existence of a profound incompatibility between the categories of continuity (wave) and discontinuity (particle) that had its origin in the constraints of human psychology and thought. He believed that, although incompatible, they are also inevitable. As one can see, this is Kierkegaard's "qualitative dialectics" once again. This cannot be surprising if one recalls Høffding's definition of Kierkegaard's influence on him—"bewitching." All this evidence, and much more that can be found in Refs. 26, 30, and 31, strengthens the independent conclusion reached here, that Bohr created complementarity under the influence of the philosophical ideas of Danish existentialism. By doing so he *freely* decided to introduce a strange new idea into physics, the idea that fundamental problems cannot be solved and that one has to live with dilemmas!

Complementarity and the uncertainty relations can be addressed by distinguishing clearly between ontological and empirical statements, which are easily confused in quantum mechanics. De Broglie, for instance, had a theory—the *double solution*[33]—that included the empirically superfluous u-wave alongside the empirically indispensable but ontologically ambiguous ψ-function one finds in quantum mechanics as well. The real wave $u = u_0 + v$ was the sum of the two empirically "incompatible" elements of the quantum world, the particle u_0 and the wave v (which was proportional to ψ). De Broglie used ψ to make empirical statements about measurements, their incompatibilities, and their statistics, and u for such ontological statements as "the particle u_0 is guided along its trajectory by the wave v according to the guidance condition $\mathbf{p} = -\nabla\varphi$." Bohm, on the other hand, gets the only wave in his theory, the confusing ψ-function, to do the guiding. De Broglie, whose *double solution* was not developed enough at the time of the 1927 Solvay conference, did away with the real wave v and ended up presenting a pilot wave theory in which the particle was guided by ψ, much as in Bohm's theory. The fact that a normalized wave of manifestly statistical character should have guided the particle caused great confusion. The theory was criticized sharply, and de Broglie soon abandoned it.

As long as measurement is not at issue, there is nothing wrong with saying that a quantum object "*is* a wave and *is* a particle." To make more empirical statements, a theory about measurements, their incompatibilities, and their statistics ought to be invoked. The most developed at the moment is quantum theory, according to which an apparatus indicating that something is a wave cannot also indicate, at the same time, that it is a particle as well. Or one may— possibly appealing to the EPR reality criterion—wish to assert that "a particle's position is contained in the infinitesimal interval $x_0 \rightarrow x_0 + dx$" and "a particle's momentum is contained in the infinitesimal interval $p_0 \rightarrow p_0 + dp$" (where x and p are conjugate). Why not? Quantum mechanics indicates, however, the incompatibility of the empirical statements "the position of a particle is measured to be within the infinitesimal interval $x_0 \rightarrow x_0 + dx$", "and, at the same time, the momentum of a particle is measured to be within the infinitesimal interval $p_0 \rightarrow p_0 + dp$." No apparatus can measure both at the same time with that precision.

REFERENCES

1. D. BOHM, *Phys. Rev.* **85**, 166–193 and 180–193 (1952).
2. L. DE BROGLIE, *Une tentative d'interprétation causale et non-linéaire de la mécanique ondulatoire*, Paris, Gauthier-Villars (1956).
3. A. EINSTEIN, in: *Albert Einstein: Philosopher-Scientist* (P. A. SCHILPP, ed.), Open Court, La Salle (1970), p. 3.
4. Quoted by MAX BORN, *Physics in My Generation*, Springer, New York (1969), p. 162.
5. A. EINSTEIN, B. PODOLSKY, and N. ROSEN, *Phys. Rev.* **47**, 777–780 (1935).
6. E. P. WIGNER, *Z. Phys.* **133**, 101–108 (1952).
7. H. A. ARAKI and M. M. YANASE, *Phys. Rev.* **20**, 622–626 (1960).
8. M. M. YANASE, *Phys. Rev.* **123**, 666–668 (1961).
9. L. W. ALVAREZ, in: *Proc. 1975 Int. Symp. Lepton and Photon Interactions at High Energies*, Stanford (1975), p. 967.
10. A. EINSTEIN, in: *Louis de Broglie, Physicien et Penseur* (A. GEORGE, ed.), Albin Michel, Paris (1953), p. 7.
11. The above derivation of Bell's inequality is a reformulation of N. D. MERMIN's. *Phys. Today* April 1985, pp. 38–47.
12. J. S. BELL, in; *Foundations of Quantum Mechanics* (B. D'ESPAGNAT, ed.), Italian Physical Society, Course IL, Academic Press, New York (1971), pp. 171–181.
13. B. D'ESPAGNAT, *Conceptions de la physique contemporaine*, Hermann, Paris (1965).
14. F. SELLERI, in; *Symposium on the Foundations of Modern Physics* (K. V. LAURIKAINEN *et al.*, eds.), Ed. Frontières, Paris (1994), pp. 255–272.
15. W. HEISENBERG, *The Physical Principles of the Quantum Theory*, University of Chicago Press (1930), p. 20.
16. J. P. WESLEY, *Classical Quantum Theory*, Benjamin Wesley, Blumberg (1996), especially Chap. 6.
17. J. R. CROCA, in: *Causality and Locality in Modern Physics* (S. JEFFERS *et al.*, eds.), Kluwer, Dordrecht (1997).
18. G. BINNIG and H. ROHRER, *Rev. Mod. Phys.* **59**, 615–625 (1987).
19. R. P. FEYNMAN, R. B. LEIGHTON, and M. SANDS, *The Feynman Lectures on Physics*, Vol. 3, Addison-Wesley, Reading (1965), pp. 1–11.
20. J. VON NEUMANN, in: *Les nouvelles theories de la physique; Conference Proceedings, Warsaw 1938*, International Institute of Intellectual Cooperation, Paris (1939), pp. 32–40.
21. N. BOHR, in: *Les nouvelles theories de la physique; Conference Proceedings, Warsaw 1938*, International Institute of Intellectual Cooperation, Paris (1939), pp. 40–41.
22. N. BOHR, *Phys. Rev.* **48**, 696–702 (1935).
23. J. VON NEUMANN, *Mathematische Grundlagen der Quantenmechanik*, Springer, Berlin (1932).
24. Quoted in A. LANDÉ, *New Foundations of Quantum Mechanics*, Cambridge University Press, Cambridge (1965), pp. 147–148.
25. G. LUKÁKS, *La distruzione della ragione*, Einaudi, Torino (1959), p. 271.
26. For a review see F. SELLERI, *Quantum Paradoxes and Physical Reality*, Kluwer, Dordrecht (1990), especially Chap. 3.
27. M. JAMMER, *The Conceptual Development of Quantum Mechanics*, Tomash Publishers, New York (1989), p. 179.
28. N. SVARTHOLM, *Søren Kierkegaard und die Moderne Physik*, University of Göteborg, preprint (1967). As far as we know the only published version of this paper is N. SVARTHOLM, in: *Che cos'è la realtá* (F. SELLERI, ed.), Jaca Book, Milano (1990), p. 113.
29. D. FAVRHOLDT, *Riv. Storia Scienza*, **2**, 445–461 (1985).

30. J. FAYE, *Niels Bohr: His Heritage and Legacy. An Anti-Realist View of Quantum Mechanics*, Kluwer, Dordrecht (1991).

31. R. N. MOREIRA, in: *Waves and Particles in Light and Matter* (A. VAN DER MERWE et al., eds.), Plenum, New York (1994), p. 395.

32. Quoted by R. N. MOREIRA, in: *Waves and Particles in Light and Matter* (A. VAN DER MERWE et al., eds.), Plenum, New York (1994), p. 397.

33. L. DE BROGLIE, *Une tentative d'interprétation causale et non-linéaire de la mécanique ondulatoire*, Paris, Gauthier-Villars (1956).

34. C. DEWDNEY and B. J. HILEY, *Found. Phys.* **12**, 27–48 (1982).

35. C. PHILIPPIDIS, C. DEWDNEY, and B. J. HILEY, *Nuoyo Cim.* **52**B, 15–28 (1979).

Chapter 3

Local Realism versus Quantum Nonlocality

The conceptual foundations of the paradox of Einstein, Podolsky, and Rosen will now be examined in greater detail. The propagation of quantum waves in configuration space gives rise to interference effects that are incompatible with local realism. If the point representing the positions of the particles making up a quantum system does indeed describe a trajectory in configuration space—guided according to the nonlocal formula $p = -\nabla\varphi$—it will be subject to a nonlocal quantum potential. There are many ways, often involving inequalities, of characterizing the nonlocal interference effects deriving from the configuration space description. One can distinguish between weak inequalities, deduced from local realism alone and never violated experimentally, and strong inequalities, which are easier to violate because they depend on further assumptions regarding detection. The more general probabilistic treatment, which rests on a generalization of the deterministic criterion used by Einstein, Podolsky, and Rosen for the identification of elements of reality, will also be dealt with.

3.1. NONLOCALITY, INTERFERENCE, QUANTUM POTENTIAL

The origins of the EPR paradox can be found as far back as Schrödinger's 1926 papers on wave mechanics, in which quantum waves were made to propagate in configuration space. Bell's inequality is, in a sense, violated by interference terms associated with such waves, which can "entangle" particles no matter how far apart they are. Again, the interference does not take place locally, in ordinary three-dimensional space, but in abstract tensor product or configuration spaces.

3.1.1. The Quantum Potential in Configuration Space

The matter of whether violations of Bell's inequality can carry a signal was discussed in 1978 by Bohm and Hiley,[1] who commented: "If it can, we will be led to a violation of the principle of Einstein's theory of relativity, because the instantaneous interaction implied by the quantum potential will lead to the possibility of a signal that is faster than light."

The whole quantum-mechanical treatment of distant systems does contain this difficulty, at least in principle. Bohm and Hiley showed that the two-body Schrödinger equation

$$i\hbar \frac{\partial \psi}{\partial t} = -\frac{\hbar^2}{2m}[\nabla_1^2 + \nabla_2^2]\psi + V\psi \tag{1}$$

for the wave function $\psi(x_1 x_2 t) = R(x_1 x_2 t)\exp[iS(x_1 x_2 t)/\hbar]$ can be written

$$\frac{\partial R^2}{\partial t} + \nabla_1 \cdot \left(R^2 \frac{\nabla_1 S}{m}\right) + \nabla_2 \cdot \left(R^2 \frac{\nabla_2 S}{m}\right) = 0 \tag{2}$$

and

$$\frac{\partial S}{\partial t} + \frac{1}{m}(\nabla_1 S)^2 + \frac{1}{m}(\nabla_2 S)^2 + V(x_1 x_2 t) + V_q(x_1 x_2 t) = 0 \tag{3}$$

where R^2 is the probability density, $V(x_1 x_2 t)$ is the external and relative potential of the two particles, and

$$V_q = -\frac{\hbar^2}{2m}\left(\frac{\nabla_1^2 R}{R} + \frac{\nabla_2^2 R}{R}\right) \tag{4}$$

∇_1 and ∇_2 are the gradient operators for particles 1 and 2. Now Eq (2) evidently describes the conservation of probability in the configuration space of the two particles. Equation (3) is a Hamilton–Jacobi equation for the system of two particles, acted on not only by the classical potential V, but also by the quantum potential $V_q(x_1 x_2 t)$.

The latter potential has peculiar nonlocal properties, since

a. It cannot be expressed as a universal function of the coordinates, as with usual potentials, because it has, in general, a different spatial dependence for different functions R.
b. It depends on $\psi(x_1 x_2 t)$ and therefore on the quantum system as a whole; since $R = R(x_1 x_2 t)$, the force acting on particle 1 depends on the instantaneous position of particle 2, and *vice versa*.
c. It does not, in general, produce a vanishing interaction between the two particles when $|x_1 - x_2| \to \infty$.

Nonlocality clearly disappears for product states. Indeed from $\psi = \psi_1 \psi_2$ it follows that

$$R(x_1 x_2 t) = R_1(x_1)R_2(x_2)$$

and the quantum potential (4) becomes

$$V_q = -\frac{\hbar^2}{2m} \left(\frac{\nabla_1^2 R_1(x_1)}{R_1(x_1)} + \frac{\nabla_2^2 R_2(x_2)}{R_2(x_2)} \right)$$

so that each particle is now acted on by a force depending only on its own position.

Bohm and Hiley embrace the nonlocal effects that, in their view, constitute the essential new quality implied by quantum theory and try to develop a physical picture of the world based on the notion of "unbroken wholeness", which they attribute to correlated quantum systems.

Nevertheless, the problem of reconciling nonlocal effects with relativity remains unsolved. In another paper Hiley[2] quotes the following opinion, expressed in 1972 by Dirac: "It [nonlocality] is against the spirit of relativity, but is the best we can do at the present time . . . and, of course, one is not satisfied with such a theory. I think one ought to say that the problem of reconciling quantum theory with relativity is not solved." Incidentally, during a 175 conference in Australia, Dirac even expressed the view that Einstein's view might ultimately prove to be correct:

> There are great difficulties . . . in connection with the present quantum mechanics. It is the best that one can do up till now. But, one should not suppose that it will survive indefinitely into the future. And I think that it is quite likely that at some future time we may get an improved quantum mechanics in which there will be a return to determinism and which will, therefore, justify the Einstein point of view.[3]

3.1.2. Quantum Correlations and Interference

The paradoxical aspects of quantum correlations are due to interference. Consider a mixture of factorizable state vectors for which the state vector $|\psi_l\rangle|\varphi_l\rangle$ has probability ρ_l. The correlation function for the dichotomic observables $A(a)$ and $B(b)$ of the subsystems α and β, respectively, is given by

$$P(ab) = \sum_l \rho_l A(al) B(bl) \tag{5}$$

where

$$A(al) = \langle \psi_l | A(a) | \psi_l \rangle, \qquad B(bl) = \langle \varphi_l | B(b) | \varphi_l \rangle$$

Since the weights ρ_l are nonnegative and sum to 1, Eq. (5) is similar to the hidden-variable expression

$$P(ab) = \int d\lambda \, \rho(\lambda) A(a\lambda) B(b\lambda) \tag{6}$$

and leads to Bell's inequality.

Take, instead, a nonfactorizable state vector $|\eta\rangle$, expressed in the polar form

$$|\eta\rangle = \sum_l \sqrt{\rho_l} |\psi_l\rangle |\varphi_l\rangle$$

The correlation function is now

$$P(ab) = \sum_{lm} \sqrt{\rho_l \rho_m} A(alm) B(blm) \qquad (7)$$

where

$$A(alm) = \langle \psi_l | A(a) | \psi_m \rangle, \qquad B(blm) = \langle \varphi_l | B(b) | \varphi_m \rangle$$

It is well known that Eq. (7) violates Bell's inequality. It is remarkable that (7) is, in many ways, similar to (6): if one considers l and m hidden variables from the purely formal point of view, one concludes that locality is satisfied [$A(a)$ does not depend on b, $B(b)$ does not depend on a, the quantity $\sqrt{\rho_l \rho_m}$, which could be thought to represent a density function, does not depend on a, b]. The reason for which (7) violates Bell's inequality is that the "density function" is not normalized to unity:

$$\sum_{lm} \sqrt{\rho_l \rho_m} = \sum_{lm} \rho_l + \sum_{l \neq m} \sqrt{\rho_l \rho_m} = 1 + \sum_{l \neq m} \sqrt{\rho_l \rho_m}$$

which is, in general, larger than unity because of the presence of interference terms (those having $l \neq m$). Violations of Bell's inequality are, therefore, a typical quantum phenomenon, since they arise from interference.

3.2. FACTORIZABLE AND NONFACTORIZABLE STATE VECTORS

The distance between a nonfactorizable state vector $|\eta\rangle$ and the closest factorizable vector is proposed as a measure of how entangled $|\eta\rangle$ is. It is then shown that factorizable state vectors cannot violate Bell's inequality and that every nonfactorizable state vector can be made to violate a Bell inequality derived from the principles of local realism.

3.2.1. Degree of Entanglement

Let us again expand the unit vector $|\eta\rangle$ with respect to the orthonormal sets $\{|\psi_i\rangle\}$ and $\{|\varphi_j\rangle\}$ that give it the polar form

$$|\eta\rangle = \sum_i c_i |\psi_i\rangle |\varphi_i\rangle$$

Let us assume that at least two of the coefficients c_i do not vanish, so it is impossible to write $|\eta\rangle$ as a product. Form the projection operator

$$P_\eta = |\eta\rangle\langle\eta|$$

which is Hermitian and, hence, in principle, represents an observable. Clearly the average value of P_η for the state represented by $|\eta\rangle$ is equal to 1.

One can then wonder by how little the average value

$$\langle x|P_\eta|x\rangle = |\langle x|\eta\rangle|^2 \tag{8}$$

can differ from 1, if the vector $|x\rangle$ is a product $|u\rangle|v\rangle$. To establish this we must find the maximum of (8).

To begin with,

$$|\langle x|\eta\rangle|^2 = \left|\left\langle uv\left|\sum_i c_i\psi_i\varphi_i\right\rangle\right|^2 = \left|\sum_i c_i\langle u|\psi_i\rangle\langle v|\varphi_i\rangle\right|^2$$

Since the coefficients are complex, phase differences will make this average value satisfy

$$|\langle x|\eta\rangle|^2 \leq \left(\sum_i |c_i|y_iz_i\right)^2 = Q^2 \tag{9}$$

where Q is a positive number defined otherwise by (9) itself,

$$y_i = |\langle u|\psi_i\rangle|, \qquad z_i = |\langle v|\varphi_i\rangle|$$

and

$$\sum_i y_i^2 = \sum_i z_i^2 = 1 \tag{10}$$

(by Parseval's relation, since the $|\psi_i\rangle$'s and $|\varphi_i\rangle$'s form bases and $|u\rangle$ and $|v\rangle$ are normalized). Now since (9) is satisfied, one can look for the maximum value of

$$\left(\sum_i |c_i|y_iz_i\right)^2 \tag{11}$$

given the restrictions (10), or since the values of y_i and z_i that maximize (11) will also maximize

$$Q = \sum_i |c_i|y_iz_i$$

one can use Q instead. A positive real-valued function reaches its maximum at the point its square reaches its maximum as well.

At its maximum, Q will be insensitive to a small variation $y_k \delta z_k + z_k \delta y_k$ in $y_i z_i$, i.e.,

$$\delta Q = \sum_k |c_k|(y_k \delta z_k + z_k \delta y_k) = 0 \tag{12}$$

Such variations must obey the restrictions (10), so

$$\sum_k y_k \delta y_k = 0 \tag{13}$$

and

$$\sum_k z_k \delta z_k = 0 \tag{14}$$

One can then add (12), (13), and (14), respectively multiplied by 1, and by the Lagrange multipliers λ_1 and λ_2. This yields

$$\sum_k (|c_k| y_k \delta z_k + |c_k| z_k \delta y_k + \lambda_1 y_k \delta y_k + \lambda_2 z_k \delta z_k)$$
$$= \sum_k [(|c_k| z_k + \lambda_1 y_k) \delta y_k + (|c_k| y_k + \lambda_2 z_k) \delta z_k] = 0$$

The factors multiplying the δy_k's and δz_k's must all vanish if we regard all these variations as independent (which is in keeping with Lagrange's approach):

$$|c_k| z_k + \lambda_1 y_k = 0 \qquad \text{(for all } k)$$
$$|c_k| y_k + \lambda_2 z_k = 0 \qquad \text{(for all } k). \tag{15}$$

Multiplying the first by y_k, the second by z_k, and subtracting the results, one has

$$\lambda_1 y_k^2 = \lambda_2 z_k^2 \qquad \text{(for all } k)$$

whence

$$\lambda_1 \sum_k y_k^2 = \lambda_2 \sum_k z_k^2$$

This means that λ_1 and λ_2 must be equal, since it is already known, from (10), that the other two factors are equal. So one can change (15) accordingly, multiply the first equation by z_k, the second by y_k, and subtract, with the result

$$|c_k|(y_k^2 - z_k^2) = 0 \qquad \text{(for all } k)$$

This means that $y_k = z_k$, for all k such that $|c_k| \neq 0$. Knowing this, one can rewrite either one of Eqs. (15):

$$|c_k| y_k + \lambda_1 y_k = 0 \qquad \text{(for all } k \text{ such that } |c_k| \neq 0)$$

which has as solutions either $y_k = 0$ or $\lambda_1 = -|c_k|$. The parameter λ_1 can assume this value for those k (in the set I_M of the values of k) for which $|c_k|$ assumes its maximum M (and it shall be demonstrated presently that the maximum has to be taken). Therefore one can choose the solution

$$k \in I_M \qquad \lambda_1 = -M$$
$$k \notin I_M \qquad y_k = 0$$

Hence,

$$\sum_{k \in I_M} y_k^2 = 1$$

and above all

$$Q = \sum_{i \in I_M} |c_i| y_i^2 = M \sum_{i \in I_M} y_i^2 = M \tag{16}$$

The largest value that Q can reach will be obtained by taking

$$M = \max_k |c_k| \tag{17}$$

Notice that, as $|\eta\rangle$ is normalized,

$$M < 1 \tag{18}$$

if more than one k exists for which $|c_k| \neq 0$ (namely if the state vector is not factorizable). Therefore

$$\langle x|P_\eta|x\rangle \leq \left(\max_k |c_k|\right)^2 < 1 \tag{19}$$

because of (8), (9), (16), and (18).

The degree of entanglement (cf. Shimony[4]) parameter ϵ can thus be defined as

$$\epsilon = 1 - \left(\max_k |c_k|\right)^2$$

and characterizes the distance between an entangled state and the closest unentangled one.

3.2.2. Bell's Inequality and Factorizable State Vectors

A remarkable property of mixtures of factorizable state vectors is that they always satisfy Bell's inequality, as first shown by Capasso, Fortunato, and Selleri.[5] Consider an ensemble E of N quantum pairs (α, β) and suppose they are described by factorizable state vectors $|\Psi_k\rangle|\Phi_k\rangle$ with frequencies n_k/N $(k = 1, 2, \ldots)$. So in the ensemble E,

$$|\Psi_1\rangle|\Phi_1\rangle \text{ applies to } n_1 \text{ pairs}$$
$$|\Psi_2\rangle|\Phi_2\rangle \text{ applies to } n_2 \text{ pairs}$$
$$\vdots \tag{20}$$
$$|\Psi_k\rangle|\Phi_k\rangle \text{ applies to } n_k \text{ pairs}$$
$$\vdots$$

and

$$\sum_k n_k = N$$

Suppose the dichotomic observables to be measured on α and β are described by operators $A(a)$ and $B(b)$, respectively, so the operator corresponding to the product of the joint measurements is $A(a) \otimes B(b)$. The correlation function predicted by quantum mechanics is then the average of the observable represented by the latter operator over the mixture (20), so

$$P(a, b) = \sum_k p_k \langle \Psi_k \Phi_k | A(a) \otimes B(b) | \Psi_k \Phi_k \rangle$$

where

$$p_k = \frac{n_k}{N}, \qquad \sum_k p_k = 1$$

The four correlation functions figuring in Bell's inequality can then be written

$$P(a, b) = \sum_k p_k \bar{A}_k \bar{B}_k$$

$$P(a, b') = \sum_k p_k \bar{A}_k \bar{B}'_k$$

$$P(a', b) = \sum_k p_k \bar{A}'_k \bar{B}_k \qquad (21)$$

$$P(a', b') = \sum_k p_k \bar{A}'_k \bar{B}'_k$$

where

$$\bar{A}_k = \langle \Psi_k | A(a) | \Psi_k \rangle \qquad \bar{B}_k = \langle \Phi_k | B(b) | \Phi_k \rangle$$
$$\bar{A}'_k = \langle \Psi_k | A(a') | \Psi_k \rangle, \qquad \bar{B}'_k = \langle \Phi_k | B(b') | \Phi_k \rangle \qquad (22)$$

These last quantities are expectation values of operators with eigenvalues ± 1. Therefore,

$$|\bar{A}_k| \leq 1; \quad |\bar{A}'_k| \leq 1; \quad |\bar{B}_k| \leq 1; \quad |\bar{B}'_k| \leq 1 \qquad (23)$$

which hold for all k.

Inserting (21) into the lhs of Bell's inequality, one obtains

$$\Delta = |P(a, b) - P(a, b') + P(a', b) + P(a', b')| \leq \sum_k p_k \Delta_k \qquad (24)$$

where

$$\Delta_k \equiv |\bar{A}_k \bar{B}_k - \bar{A}_k \bar{B}'_k| + |\bar{A}'_k \bar{B}_k + \bar{A}'_k \bar{B}'_k| \leq |\bar{B}_k - \bar{B}'_k| + |\bar{B}_k + \bar{B}'_k|$$

and hence

$$\Delta_k \leq 2 \tag{25}$$

Two real numbers x and y, such that $|x| \leq 1$ and $|y| \leq 1$, always satisfy $|x - y| + |x + y| \leq 2$.

If (25) is inserted in (24) one finally obtains

$$\Delta \leq 2$$

which is Bell's inequality.

3.2.3. Bell's Inequality and Nonfactorizable State Vectors

It is well known that Bell's (weak) inequality derived from local realism alone is contradicted by quantum-mechanical predictions for certain nonfactorizable state vectors of correlated spins or polarizations. The following question will now be considered: Is it possible to show that *any* given nonfactorizable state vector of correlated quantum systems *necessarily* implies a violation of Bell's inequality?

Consider the general form for a nonfactorizable state vector of a composite system given by

$$|\eta\rangle = \sum_{ij} f_{ij} |\xi_i\rangle |\zeta_j\rangle \tag{26}$$

where the $|\xi_i\rangle$'s and $|\zeta_j\rangle$'s denote complete orthonormal sets of eigenstates corresponding to the correlated subsystems α and β, respectively.

The unit vector $|\eta\rangle$ can be written in the form

$$|\eta\rangle = \sum_i \sqrt{\rho_i} |\nu_i\rangle |\gamma_i\rangle \tag{27}$$

where the ρ_i's are real and nonnegative (phases can always be absorbed by the basis). Now to show that, at least in principle, observables can always be chosen in such a way that Bell's inequality is violated for any state vector of the type given by Eq. (27), consider pairs of noncommuting dichotomic observables (with eigenvalues ± 1) D_S, D_S' pertaining to α and D_T, D_T' pertaining to β, with the following definitions for them:

$$
\begin{aligned}
D_S &= 2P_S - 1 & D_T &= 2P_T - 1 \\
D_S' &= 2P_S' - 1 & D_T' &= 2P_T' - 1
\end{aligned}
\tag{28}
$$

where the projection operators

$$
\begin{aligned}
P_S &= |\nu_1\rangle\langle\nu_1| \\
P_T &= |\gamma_1\rangle\langle\gamma_1| \\
P_S' &= [\alpha_1|\nu_1\rangle + \alpha_2|\nu_2\rangle][\alpha_1^*\langle\nu_1| + \alpha_2^*\langle\nu_2|] \\
P_T' &= [\beta_1|\gamma_1\rangle + \beta_2|\gamma_2\rangle][\beta_1^*\langle\gamma_1| + \beta_2^*\langle\gamma_2|]
\end{aligned}
\tag{29}
$$

and where $|\alpha_1|^2 + |\alpha_2|^2 = 1 = |\beta_1|^2 + |\beta_1|^2$. Note that, for simplicity, the observables above have been defined on a two-dimensional subspace of the larger Hilbert space containing $|\eta\rangle$, but this is enough. Assume that both ρ_1 and ρ_2 are nonzero, which can be done without loss of generality, since in (27) there are certainly at least two coefficients $\rho_j \neq 0$.

Using Eqs. (27), (28), and (29) it is straightforward algebra to obtain the following results:

$$\langle \eta | D_S \otimes D_T | \eta \rangle = 1$$
$$\langle \eta | D_S' \otimes D_T | \eta \rangle = (1 - \Sigma) + \Sigma \Delta \alpha$$
$$\langle \eta | D_S \otimes D_T' | \eta \rangle = (1 - \Sigma) + \Sigma \Delta \beta \tag{30}$$
$$\langle \eta | D_S' \otimes D_T' | \eta \rangle = (1 - \Sigma) + \Sigma \Delta \alpha \Delta \beta$$
$$+ \sqrt{(\Sigma^2 - \Delta \rho^2)(1 - \Delta \alpha^2)(1 - \Delta \beta^2)} \cos(\varphi_\alpha + \varphi_\beta)$$

where

$$\Sigma = \rho_1 + \rho_2 \qquad \Delta \alpha = |\alpha_1|^2 - |\alpha_2|^2$$
$$\Delta \rho = \rho_1 - \rho_2 \qquad \Delta \beta = |\beta_1|^2 - |\beta_2|^2 \tag{31}$$

and φ_α, φ_β are the relative phases of α_1, α_2 and of β_1 and β_2, respectively.

Recall that the standard form of Bell's inequality in the present notation is

$$-2 \leq \langle D_S \otimes D_T \rangle - \langle D_S' \otimes D_T \rangle + \langle D_S \otimes D_T' \rangle + \langle D_S' \otimes D_T' \rangle \leq 2 \tag{32}$$

A condition for the violation of Bell's inequality is

$$1 - (1 - \Sigma) - \Sigma \Delta \alpha + (1 - \Sigma) + \Sigma \Delta \beta + (1 - \Sigma) + \Sigma \Delta \alpha \Delta \beta$$
$$+ \sqrt{(\Sigma^2 - \Delta \rho^2)(1 - \Delta \alpha^2)(1 - \Delta \beta^2)} \cos(\varphi_\alpha + \varphi_\beta) > 2 \tag{33}$$

and reduces to

$$\cos(\varphi_\alpha + \varphi_\beta) > \frac{\Sigma}{\sqrt{\Sigma^2 - \Delta \rho^2}} \frac{1 + \Delta \alpha}{\sqrt{1 - \Delta \alpha^2}} \frac{1 - \Delta \beta}{\sqrt{1 - \Delta \beta^2}} \tag{34}$$

according to Eqs. (30)–(32).

To show that condition (34) can be satisfied by an appropriate choice of observables, assume that φ_α and φ_β are chosen such that

$$\cos(\varphi_\alpha + \varphi_\beta) = 1 \tag{35}$$

which is clearly possible since P_S' and P_T', and hence φ_α and φ_β, are arbitrary. Then the square of (34) reduces to the form

$$1 - \frac{\Delta \rho^2}{\Sigma^2} > \frac{1 + \Delta \alpha}{1 - \Delta \alpha} \frac{1 - \Delta \beta}{1 + \Delta \beta} \tag{36}$$

Rememberiing that $\Delta\rho^2 < \Sigma^2$ because in (31) both ρ_1 and ρ_2 are positive and nonzero, it is evident that condition (36) can be easily satisfied by choosing $\Delta\alpha$ and $\Delta\beta$ appropriately within the intervals $-1 \leq \Delta\alpha \leq 1$, $-1 \leq \Delta\beta \leq 1$. For example, if $\Delta\rho^2 = 0$, take $-\Delta\alpha = \Delta\beta = 0.5$, but if $\Delta\rho^2 \neq 0$ take $\Delta\alpha = 0$ and $\Delta\beta = \Delta\rho^2/\Sigma^2$. In the latter case (36) becomes

$$1 - \frac{\Delta\rho^2}{\Sigma^2} > \frac{1 - \Delta\rho^2/\Sigma^2}{1 + \Delta\rho^2/\Sigma^2}$$

which is clearly satisfied.

This completes the proof of the theorem whose statement can be formulated as follows. For any given nonfactorizable state vector of correlated quantum systems it is always possible to choose observables in such a way that Bell's inequality is violated by quantum-mechanical predictions.

Recalling that factorizable state vectors necessarily satisfy Bell's inequality, it can be concluded that nonfactorizable state vectors are necessary and sufficient for the violation of Bell's inequality. This result reinforces the notion that the incompatibility with local realism is rooted in the use of nonfactorizable state vectors in quantum mechanics, and though Bell's inequality does not contain all the restrictions implied by local realism (see Section 3.5) it is nevertheless sufficient for the purpose of displaying the incompatibility between quantum theory and local realism with great generality.

The question as to which nonfactorizable state vectors are actually realizable and which are the relevant observables that can be found in practice for testing unambiguously the quantum-mechanically predicted violation of Bell's inequality is, of course, of crucial importance. Since many of the nonfactorizable state vectors are consequences of fundamental principles used in quantum mechanics (such as the conservation of angular momentum, invariance conditions like those of parity and charge conjugation, symmetry properties of wave functions of bosons and fermions), the general theorem just proved provides theoretical support for studies in quest of suitable examples to discriminate between quantum mechanics and local realism using Bell's inequalities.

Another form of the partial compatibility of quantum mechanics with local realism has been found by Landau.[6] Consider the operator corresponding to Bell's observable:

$$\Gamma = D_S(D_T - D_{T'}) + D_{S'}(D_T + D_{T'}) \tag{37}$$

Notice that such observables can always be written in terms of projection operators as in (28) where

$$P_S^2 = P_S, \qquad P_{S'}^2 = P_{S'}$$
$$P_T^2 = P_T, \qquad P_{T'}^2 = P_{T'}$$

so that

$$D_S^2 = D_{S'}^2 = D_T^2 = D_{T'}^2 = I \tag{38}$$

For example,

$$D_S^2 = (2P_S - I)^2 = 4P_S^2 - 4P_S + I = I$$

Using (38) it is easy to show that

$$\Gamma^2 = 4 + [D_S, D_{S'}][D_T, D_{T'}]$$

Now suppose that

$$[D_S, D_{S'}] = 0 \qquad \text{and/or} \qquad [D_T, D_{T'}] = 0 \tag{39}$$

In such cases $\Gamma^2 = 4$. Let $|\chi_i\rangle$ be an eigenvector of Γ with eigenvalue γ_i. Clearly

$$\Gamma^2 |\chi_i\rangle = \gamma_i^2 |\chi_i\rangle$$

But $\Gamma^2 = 4$ when (39) is satisfied, so $\gamma_i^2 = 4$, and therefore $\gamma_i = \pm 2$. Thus, the eigenvalues of Γ are ± 2. Expanding a general vector $|\psi\rangle$ with respect to a complete orthonormal set χ_i of eigenvectors of Γ, we have

$$\langle \psi | \Gamma | \psi \rangle = \sum_{ij} c_i^* c_j \langle \chi_i | \Gamma | \chi_j \rangle = \sum_{ij} c_i^* c_j \gamma_j \langle \chi_i | \chi_j \rangle$$

$$= \sum_i |c_i|^2 \gamma_i \leq 2 \sum_i |c_i|^2 = 2$$

Therefore Bell's inequality cannot be violated when the dichotomic observables satisfy (39).

3.3. PROBABILISTIC LOCAL REALISM

In this section the probabilistic foundations of local realism are developed. In place of the EPR reality criterion one has to introduce a probabilistic reality criterion that attributes real physical properties to statistical ensembles for which probabilities can be predicted.[7] These "real physical properties" are similar to the propensities of Popper's realism.[8] When the locality assumption is added, one obtains a very general formulation of local realism that leads, for example, to a new proof of Bell's theorem.

3.3.1. Propensities and Probabilities

Let S and T be two sets of N objects of the same kind (photons, neutrons, kaons, ...):

$$S = \{\alpha_1, \alpha_2, \ldots, \alpha_N\}, \qquad T = \{\beta_1, \beta_2, \ldots, \beta_N\} \tag{40}$$

The objects are produced in pairs: α_1 with β_1, α_2 with $\beta_2, \ldots, \alpha_N$ with β_N. The different pairs are totally independent of one another. Measurements of the dichotomic physical quantities $A(a) = \pm 1$ and $B(b) = \pm 1$ (a and b are experimental parameters) are performed on the α_i's of S and on the β_i's of T, respectively, when they are far apart.

Subsets of, say, S characterized by certain properties can be considered, such as $\{\alpha_i : A(a) = \mu\}$, where $A(a) = \mu$ means that a measurement of $A(a)$ on α_i gives the result $\mu = \pm 1$. Since the objects are produced in pairs, subsets of S can also be defined with reference to T, so we could consider $\{\alpha_i : B_i(b) = \nu\}$, for instance. Frequencies such as

$$f(a_\pm) = \frac{1}{N} \, \mathrm{card}\{\alpha_i : A_i(a) = \mu\} \tag{41}$$

($\mathrm{card}\{\xi\}$ is the number of elements in the set $\{\xi\}$) can be defined as well.

Einstein, Podolsky, and Rosen identified an *element of reality* in a single quantum system when the result of a measurement could be predicted with certainty, without disturbing the system. Because one is here considering the more general case in which this certainty may not be achievable, something weaker than an element of reality is needed. This is the *propensity*, to identify which the following probabilistic reality criterion (PRC) shall be adopted. This is the natural generalization of the criterion used by Einstein, Podolsky, and Rosen:

Probabilistic Reality Criterion. Given a set S of N α-particles, if one can

1. Predict that measurements of $A(a)$ on the α_i's of a subset S' of S will give the results $+1$ and -1 with the probabilities (frequencies) $f(a_+)$ and $f(a_-)$, respectively
2. Predict the population N' of S' ($0 < N' \leq N$)
3. Make all these predictions without disturbing the α-objects of S and S' in any way

then an objective, real propensity λ_a belonging to S' is assumed to fix the probabilities:

$$f(a_+) = f(a_+, \lambda_a), \qquad f(a_-) = f(a_-, \lambda_a) \tag{42}$$

The propensities and the corresponding frequencies associated with subsets of S will generally be different from those of the entire set.

Popper considered competing propensities for different outcomes of a process (e.g., of a measurement). Here he would probably have introduced two different propensities $\lambda_{a\pm}$ for the two outcomes $A(a) = \pm 1$, but this is not inconsistent with the use of the symbol λ_a, which can be written

$$\lambda_a = \{\lambda_{a+}, \lambda_{a-}\}$$

as the set of both propensities. Hence, one can view the frequency $f(a_+)$ of the outcome $A(a) = +1$, for instance, as the consequence of both propensities present in S' and represent this as before with Eqs. (42).

This PRC should not be viewed as an attempt to "define" propensities. Their existence is being postulated and a sufficient, but not necessary, condition for their identification given. That is, propensities could exist without being capable of identification by means of the criterion. More or less the same applies to the deterministic criterion of Einstein, Podolsky, and Rosen. This PRC, then, will be applied to the subsets $S(b_\pm)$.

Consider an EPR experiment in which there are two observers: O_α, who measures the dichotomic observables $A(a)$ on the α-particles of S, and O_β, who measures the dichotomic observable $B(b)$ on the β-particles of T (Fig. 3.1). Measuring instruments are assumed to be perfectly efficient: every single act of measurement produces one of the two outcomes ± 1. Now O_β's measurement on β_i, which we can assume to take place (in the laboratory frame) before O_α's measurement on α_i, will split T into the two subsets

$$T(b_\pm) = \{\beta_i \colon B_i(b) = \pm 1\}.$$

This division of T will produce corresponding divisions

$$S(b_\pm) = \{\alpha_i \colon B_i(b) = \pm 1\}$$

of S and

$$E(b_\pm) = \{(\alpha_i, \beta_i) \colon B_i(b) = \pm 1\}$$

of the set

$$E = \{(\alpha_1, \beta_1), (\alpha_2, \beta_2), \ldots, (\alpha_N, \beta_N)\}$$

of (α_i, β_i) pairs.

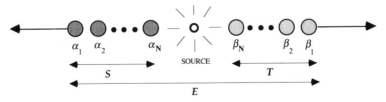

FIGURE 3.1. The source emits α and β particles simultaneously and in opposite directions. First the (α_1, β_1) pair is emitted; then the (α_2, β_2) pair; ... finally the (α_N, β_N) pair. The set S contains all the α's; the set T contains all the β's; the set E contains all the (α, β) pairs.

It will be convenient to use $N(b_\pm)$ to represent the populations of the sets $T(b_\pm)$, $S(b_\pm)$, and $E(b_\pm)$; i.e.,

$$N(b_\pm) = \text{card}\{\beta_i : B_i(b) = \pm 1\} = \text{card}\{\alpha_i : B_i(b) = \pm 1\}$$

$$= \text{card}\{(\alpha_i, \beta_i) : B_i(b) = \pm 1\}$$

Since the observable $B(b)$ is dichotomic,

$$N(b_+) + N(b_-) = N \tag{43}$$

If the probabilities $g(b_\pm)$ of measuring $B(b)$ in T are identified with the frequencies introduced quite generally by (41), it follows that

$$g(b_\pm) = \left(\frac{1}{N}\right) N(b_\pm)$$

A knowledge of previous experiments enables O_β to predict that O_α will subsequently find $A(a) = \pm 1$ with frequencies $f(a_\pm | b_+)$ in $S(b_+)$, and $A(a) = \pm 1$ with frequencies $f(a_\pm | b_-)$ in $S(b_-)$. Again, $f(a_\pm | b_+)$ in $S(b_+)$ and $f(a_\pm | b_-)$ in $S(b_-)$ are, in general, different from the probabilities $f(a_\pm)$ of finding $A(a) = \pm 1$ in the whole set S. Observer O_α, finding the very frequencies predicted by O_β, concludes that these frequencies are generated by a real feature of $S(b_\pm)$, namely a propensity. One can write

$$f(a_\pm | b_+) = f[a_\pm, \lambda_a(b_+)]$$
$$f(a_\pm | b_-) = f[a_\pm, \lambda_a(b_-)] \tag{44}$$

to indicate the dependence of the frequencies $f(a_\pm)$ on the propensities $\lambda_a(b_\pm)$ of the subsets $S(b_\pm)$. Now the dependence of $\lambda_a(b_\pm)$ on the splitting of T is established at the birth of the pairs. The notation $\lambda_a(b_\pm)$ only means that the sets $S(b_\pm)$ and their propensities are defined by the (potential) measurement of $B(b)$ and is not meant to imply the existence of a distant coupling of T and S capable of modifying λ_a. Indeed the following locality assumption can be added to the PRC:

Locality Assumption. Measurements performed on the β-objects of T do not in any way generate or modify the propensities of the α-objects of S, and *vice versa*.

An "arrow of time" assumption could be added at this point, but it is best not to insist on this here. The propensities $\lambda_a(b_\pm)$ are therefore assumed to be real properties of subsets $S(b_\pm)$ of S even if no measurement of $B(b)$ is performed. Not to admit this would imply that the propensities for the α-particles are created via some action-at-a-distance mechanism by the measurements on the β-particles, which is precisely what the locality assumption excludes.

3.3.2. Quantum Sets Are Not Homogeneous

Other probabilities can similarly be introduced for a different splitting of E into $E(b'_+)$ and $E(b'_-)$ arising from actual (or possible) measurements of $B(b')$ on the β-particles. Naturally S also splits into $S(b'_+)$ and $S(b'_-)$. Considering once more the observable $A(a)$ of the α-particles, the PRC can be applied to the case at hand; hence,

$$f(a_\pm | b'_+) = f[a_\pm, \lambda_a(b'_+)]$$
$$f(a_\pm | b'_-) = f[a_\pm, \lambda_a(b'_-)] \tag{45}$$

where the first frequency holds for $S(b'_+)$ and the second for $S(b'_-)$. Different subsets generally have different probabilities, and $\lambda_a(b'_+)$ will thus be different from $\lambda_a(b_+)$ because different effects (probabilities) imply different causes (propensities). Clearly

$$S = S(b_+) \cup S(b_-) \tag{46}$$

and therefore $S(b'_+)$, a subset of S, will consist of elements of $S(b_+)$ and $S(b_-)$. By virtue of (46),

$$S(b'_+) = S(b'_+) \cap S = S(b'_+) \cap [S(b_+) \cup S(b_-)]$$

and since $S(b_+)$ and $S(b_-)$ are disjoint it follows

$$S(b'_+) = [S(b'_+) \cap S(b_+)] \cup [S(b'_+) \cap S(b_-)] \tag{47}$$

It can be demonstrated by *reductio ad absurdum* that these sets are generally not homogeneous; i.e., they may have subsets with different propensities and frequencies. Suppose $S(b_+)$ and $S(b_-)$ are homogeneous, which means their frequencies for the outcome $A(a) = +1$ apply to their subsets $S(b'_+) \cap S(b_+)$ and $S(b'_+) \cap S(b_-)$ as well. Suppose, furthermore, that a fraction γ $(0 \leq \gamma \leq 1)$ of the elements of $S(b'_+)$ are in $S(b'_+) \cap S(b_+)$, with $1 - \gamma$ in $S(b'_+) \cap S(b_-)$. The

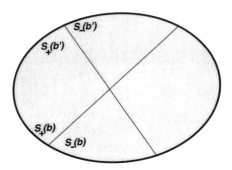

FIGURE 3.2. Graphic representation of the set S of α particles (large oval) and of its splitting into two subsets. $S_\pm(b)$ $[S_\pm(b')]$ arise from the possible measurements of $B(b)$ $[B(b')]$ on the correlated β's.

frequency of the outcome $+1$ for $S(b'_+)$ is the weighted average of the corresponding frequencies associated with the disjoint subsets that make it up. Here these frequencies are assumed to be the same as those of sets *of which they are proper subsets*; i.e., they are assumed to be

$$f[a_+, \lambda_a(b_+)] \qquad \text{and} \qquad f[a_+, \lambda_a(b_-)] \qquad (48)$$

which are the frequencies of $S(b_+)$ and $S(b_-)$. Were those the right frequencies, it would follow that

$$f[a_+, \lambda_a(b'_+)] = \gamma f[a_+, \lambda_a(b_+)] + (1 - \gamma) f[a_+, \lambda_a(b_-)]$$

so $f[a_+, \lambda_a(b'_+)]$ would lie in the interval from $f[a_+, \lambda_a(b_+)]$ to $f[a_+, \lambda_a(b_-)]$.
The same can be done with $S(b'_-)$, which can also be represented as a union

$$S(b'_-) = [S(b'_-) \cap S(b_+)] \cup [S(b'_-) \cap S(b_-)]$$

of subsets of $S(b_+)$ and $S(b_-)$, which subsets can again be assumed to have the same frequencies (48) as $S(b_+)$ and $S(b_-)$. The frequency of the outcome $A(a) = +1$ for $S(b'_-)$ would then, much as before, be a weighted average of the frequencies of the subsets of which it is the union:

$$f[a_+, \lambda_a(b'_-)] = \gamma' f[a_+, \lambda_a(b_+)] + (1 - \gamma') f[a_+, \lambda_a(b_-)]$$

where a fraction γ' of the elements of $S(b'_-)$ are also in $S(b'_-) \cap S(b_+)$, and a fraction $1 - \gamma'$ in $S(b'_-) \cap S(b_-)$. Hence $f[a_+, \lambda_a(b'_-)]$, too, would be bounded on either side by the same frequencies (48). But the fact that $f[a_+, \lambda_a(b'_\pm)]$ should be bounded by $f[a_+, \lambda_a(b_\pm)]$ suggests a groundless asymmetry between b and b', which are entirely arbitrary and can be interchanged. These above frequencies could *a priori* be constant, but that is certainly not always the case in nature. Therefore $S(b_\pm)$ cannot in general be homogeneous.

That frequencies $f[a_\pm, \lambda_a(b_\pm)]$ can vary with b means that they will not always be equal to those associated with the entire set S, and hence suggests in itself an inhomogeneity of S.

The foregoing considerations are general enough to apply to sets S, T, and E of any kind. To apply them to quantum systems in particular, take the quantum observables

$$A(a) = \boldsymbol{\sigma} \cdot \hat{\boldsymbol{a}} = \sum_{i=1}^{3} \sigma_i a_i, \qquad B(b) = \boldsymbol{\tau} \cdot \hat{\boldsymbol{b}} = \sum_{j=1}^{3} \tau_j b_j$$

where σ_i and τ_j $(i, j = 1, 2, 3)$ are the Pauli matrices for two spin-$\frac{1}{2}$ particles, α and β; a_i are b_j the components of the unit vectors $\hat{\boldsymbol{a}}$ and $\hat{\boldsymbol{b}}$. The dichotomic observables $A(a)$ and $B(b)$, which have eigenvalues ± 1, represent the spin components of particles α and β along the directions a and b. The vector describing the singlet state can be written

$$\eta_s = \frac{1}{\sqrt{2}}[u(b_+)v(b_-) - u(b_-)v(b_+)]$$

where

$$(\boldsymbol{\sigma} \cdot \hat{\boldsymbol{b}})u(b_\pm) = \pm u(b_\pm), \qquad (\boldsymbol{\tau} \cdot \hat{\boldsymbol{b}})v(b_\pm) = \pm v(b_\pm)$$

Since it is rotationally invariant, η_s will be an eigenvector of $\boldsymbol{\sigma} \cdot \hat{\boldsymbol{b}} + \boldsymbol{\tau} \cdot \hat{\boldsymbol{b}}$ belonging to the eigenvalue 0 regardless of what $\hat{\boldsymbol{b}}$ is.

Suppose next that $\boldsymbol{\tau} \cdot \hat{\boldsymbol{b}}$ is measured on the set T of β particles: according to quantum theory, the results ± 1 will be found with the same frequency $\frac{1}{2}$. In the notation of Eq. (43),

$$N(b_+) = N(b_-) = N/2$$

These populations $N(b_\pm)$ refer not only to the subsets $T(b_\pm)$ but also to $E(b_\pm)$ and $S(b_\pm)$. According to quantum theory, the "reduction of the state vector" takes place during measurement and the new states

$$u(b_-)v(b_+) \quad \text{in } E(b_+), \qquad u(b_+)v(b_0) \quad \text{in } E(b_-) \tag{49}$$

must be considered. It is now easy to calculate the quantum-mechanically predicted probabilities $f(a_\pm|b_\pm)$ of the last section. For example,

$$f_{QM}(a_+|b_+) = \langle u(b_-)v(b_+)|(1 + \boldsymbol{\sigma} \cdot \hat{\boldsymbol{a}})/2|u(b_-)v(b_+)\rangle = \sin^2\left(\frac{\hat{a} - \hat{b}}{2}\right)$$

where $a - b$ is the angle between the directions a and b. More generally,

$$f_{QM}(a_+|b_+) = f_{QM}(a_-|b_-) = \sin^2\left(\frac{\hat{a} - \hat{b}}{2}\right)$$

$$f_{QM}(a_+|b_-) = f_{QM}(a_-|b_+) = \cos^2\left(\frac{\hat{a} - \hat{b}}{2}\right)$$

The main point deduced in the last section from probabilistic local realism is that the conditional probabilities must emerge as necessary consequences of objectively existing propensities. Of course propensities are always there, but probabilities become real frequencies only when they are measured. From Eq. (44),

$$f_{lr}(a_+|b_+) = f[a_+, \lambda_a(b_+)] = F(\hat{a} - \hat{b})$$
$$f_{lr}(a_+|b_-) = f[a_+, \lambda_a(b_-)] = 1 - F(\hat{a} - \hat{b}) \tag{50}$$

where F is the (as yet unknown) prediction of probabilistic local realism and the dependence on $\hat{a} - \hat{b}$ can be justified on the basis of rotational invariance. Bell's theorem, whose validity in the present context will be shown later, does not allow F to equal the quantum-mechanical prediction. It follows from Eq. (45) that

$$f_{lr}(a_+|b'_+) = f[a_+, \lambda_a(b'_+)] = F(\hat{a} - \hat{b}')$$
$$f_{lr}(a_+|b'_-) = f[a_+, \lambda_a(b'_-)] = 1 - F(\hat{a} - \hat{b}')$$

(51)

Equation (47) shows that $S(b'_+)$ splits into two subsets. By assuming the homogeneity of $S(b_\pm)$, it would now follow, as a consequence of the foregoing *reductio* argument, that $F(\hat{a} - \hat{b})$ is bounded on either side by $F(\hat{a} - \hat{b})$ and $1 - F(\hat{a} - \hat{b})$. However, it will not lie in this interval for arbitrary values of a, b, b' if the predictions [Eqs. (50) and (51)], while respecting Bell's inequality, are even vaguely similar to the quantum-theoretical expressions. Therefore, the subsets $S(b_\pm)$ cannot be homogeneous. Since these subsets have been associated with the quantum-mechanical states $u(b_\pm)$ [see Eq. (49)], one must conclude that, in an EPR situation, the eigenvectors of the spin observables cannot represent homogeneous ensembles. This conclusion follows rigorously from the properties of the singlet state; a more general validity, for all states of spin-$\frac{1}{2}$ particles, can be conjectured.

This is a generalization of the incompleteness intended by Einstein, Podolsky, and Rosen. Von Neumann showed[9] that the assumption of quantum-mechanical completeness implies that pure states describe homogeneous ensembles (which means that every subset is represented by the same state). We have come instead to the conclusion that pure states must describe inhomogeneous ensembles, if our assumptions concerning realism (propensities) and locality are true in nature. Hence, *if probabilities belonging to the sets $S(b_\pm)$ for which they are predicted are consequences of local propensities, they must result from averages over other probabilities, in general different for different subsets of $S(b_\pm)$, but constant within every such subset.* These probabilities of homogeneous subsets of S and T will be dependent on local propensities and will therefore be totally independent of what is eventually measured on the other set of particles (T and S, respectively). This is a consequence of the locality assumption.

3.3.3. Homogeneous Sets and Probabilities

Recalling how S was split into $S(b_+)$ and $S(b_-)$ even without measurement of $B(b)$, it becomes clear that the same splitting applies to E, which consists of S and T (Fig. 3.3). Again, $S(b_+)$ and $S(b_-)$, as well as their union S, cannot be homogeneous with respect to the probabilities of finding $A(a) = \pm 1$; and similarly—inverting the roles of the α- and β-particles—with respect to the probabilities of finding $B(b) = \pm 1$ in T.

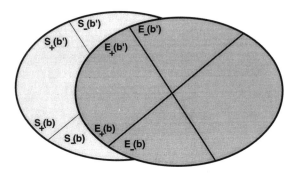

FIGURE 3.3. Graphic representation of the set E of (α, β) pairs (dark oval) and of its splitting into two subsets. $E_{\pm}(b)$ [$E_{\pm}(b')$] arise from the possible measurements of $B(b)$ [$B(b')$] on the β's. The corresponding splitting of the set S of α particles into two subsets, $S_{\pm}(b)$ and $S_{\pm}(b')$ is also shown, as in Fig. 3.2.

Recalling the results of the previous section, consider a splitting of E (arising from a similar splitting of S) into the subsets $\sigma_1(a), \sigma_2(a), \ldots, \sigma_r(a)$, where $\sigma_i(a)$ ($i = 1, 2, \ldots, r$), by definition, homogeneous for the local probabilities $f_i(a_{\pm}, \lambda_a)$ of obtaining $A(a) = \pm 1$, respectively, and λ_a is the local propensity for the same results, which has now no dependence on what is done with the β's (Fig. 3.4). Consider also a different splitting of E (arising out of a similar splitting of T) into $\tau_1(b), \tau_2(b), \ldots, \tau_s(b)$, where $\tau_j(b)$ is, by definition, homogeneous for the local probabilities $g_j(b_{\pm}, \lambda_b)$ of obtaining $B(b) = \pm 1 (j = 1, 2, \ldots, s)$, and λ_b is a local propensity, independent of what is done with the α's. Obviously, the union of these subsets must give E in both cases; that is,

$$\sigma_1(a) \cup \sigma_2(a) \cup \cdots \cup \sigma_r(a) = E$$
$$\tau_1(b) \cup \tau_2(b) \cup \cdots \cup \tau_s(b) = E \tag{52}$$

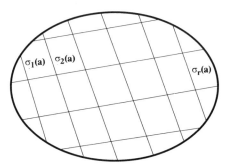

FIGURE 3.4. Graphic representation of the set E of (α, β) pairs (large oval) and of its splitting into r subsets $\sigma_1(a), \sigma_2(a), \ldots, \sigma_r(a)$ that are homogeneous for the probabilities of the results ± 1 of the observable $A(a)$ measurable on the set S of α particles.

The dependence of σ_i on a and τ_j on b reflects the fact that a subset that is homogeneous for an observable (e.g., $A(a)$] is not expected, in general, to be homogeneous for a different observable [e.g., $A(a')$].

One can, however, easily find smaller subsets that are homogeneous for two observables. For $A(a)$ and $B(b)$ they are

$$E_\lambda(a, b) = \sigma_i(a) \cap \tau_j(b)$$

where the single index λ has been chosen, for simplicity, to correspond bijectively to the pair of indices (i, j). Because i can assume r values and j can assume s, the index λ can assume rs different values. Notice that $\sigma_i(a)$ $[\tau_j(b)]$ remains homogeneous for $A(a)$ $[B(b)]$ no matter what is measured on the β- $[\alpha$-] objects. This is a consequence of the assumption of separability.

This can be generalized to an arbitrary number of observables of α and β. Considering m values $(a_1, a_2, \ldots a_m)$ of the argument of $A(a)$ and n values $(b_1, b_2, \ldots, b_n)$ of the argument of $B(b)$, the following subsets of E can be introduced that are homogeneous for the probabilities of the values ± 1 of the indicated observable:

$$\sigma_1(a_1), \sigma_2(a_1), \ldots, \sigma_{r_1}(a_1), \text{ all homogeneous for } A(a_1)$$
$$\sigma_1(a_2), \sigma_2(a_2), \ldots, \sigma_{r_2}(a_2), \text{ all homogeneous for } A(a_2) \qquad (53)$$
$$\sigma_1(a_m), \sigma_2(a_m), \ldots, \sigma_{r_m}(a_m), \text{ all homogeneous for } A(a_m)$$

and

$$\tau_1(b_1), \tau_2(b_1), \ldots, \tau_{s_1}(b_1), \text{ all homogeneous for } B(b_1)$$
$$\tau_1(b_2), \tau_2(b_2), \ldots, \tau_{s_2}(b_2), \text{ all homogeneous for } B(b_2) \qquad (54)$$
$$\tau_1(b_n), \tau_2(b_n), \ldots, \tau_{s_n}(b_n), \text{ all homogeneous for } B(b_n)$$

The homogeneity of the these sets has the usual meaning: all the subsets of $\sigma_1(a_1)$ have the same probabilities $f_1(a_1\pm, \lambda_{a_1})$ for the results $A(a_1) = \pm 1$, where λ_{a_1} is a local propensity; all the subsets of $\sigma_2(a_1)$ have the same probabilities $f_2(a_1\pm, \lambda_{a_1})$ for the results $A(a_2) = \pm 1$, and so on. Clearly the union of the sets of every line of expressions (53) and (54) always gives E, much as in Eqs. (52).

By means of suitable intersections, one can define smaller subsets in which all the observables considered have constant probability. One can write

$$E_\lambda(a_1, \ldots, a_m, b_1, \ldots, b_n) = \sigma_{i_1}(a_1) \cap \cdots \cap \sigma_{i_m}(a_m) \cap \tau_{j_1}(b_1) \cap \cdots \cap \tau_{j_n}(b_n)$$

for a typical subset homogeneous for all the probabilities of the ± 1 results of the $m + n$ observables considered. The single index λ used in the preceding equation corresponds bijectively to the set of indices $\{i_1, i_2, \ldots, i_m, j_1, j_2, \ldots, j_n\}$. The number $t(m, n)$ of such sets can easily be calculated; i_1 can be chosen in r_1

different ways, ..., i_m in r_m, j_1 in $s_1, \ldots, j_n$ in s_n, all choices being independent. Therefore

$$t(m, n) = r_1 r_2 \cdots r_m s_1 s_2 \cdots s_n$$

Notice, however, that $t(m, n)$ is in general expected to change if some arguments $a_1, \ldots, b_n$ are modified; this is clear from the notation of Eqs. (53) and (54), where s_1 [the number of subsets of E homogeneous for $B(b_1)$] has the same index as b_1, precisely because they depend on one another and a modification of b_1 can produce a change in s_1. So the set I of indices λ [set of integers from 1 to $t(m, n)$] depends on the arguments of the observables:

$$I = I(a_1, \ldots, a_m, b_1, \ldots, b_n) = I(z) \tag{55}$$

The notation has been simplified by the introduction of a "vector" z having as $m + n$ components the arguments of the observables, given by

$$z = \{a_1, \ldots, a_m, b_1, \ldots, b_n\} \tag{56}$$

The homogenous subsets and their populations can then be written

$$E_\lambda(z) = E_\lambda(a_1, \ldots, a_m, b_1, \ldots, b_n) \tag{57}$$

and

$$N_\lambda(z) = N_\lambda(a_1, \ldots, a_m, b_1, \ldots, b_n) \tag{58}$$

This simplified notation will soon prove useful.

The basic probabilities, which are all constant for pairs belonging to a subset $E_\lambda(z)$, are

$$
\begin{aligned}
&f_\lambda(a_1 \pm, \lambda_{a_1}), && \text{probabilities for } A(a_1) = \pm 1 \\
&f_\lambda(a_2 \pm, \lambda_{a_2}), && \text{probabilities for } A(a_2) = \pm 1 \\
&\cdots\cdots\cdots\cdots\cdots\cdots\cdots\cdots\cdots\cdots\cdots \\
&f_\lambda(a_m \pm, \lambda_{a_m}), && \text{probabilities for } A(a_m) = \pm 1
\end{aligned}
\tag{59}
$$

and

$$
\begin{aligned}
&g_\lambda(b_1 \pm, \lambda_{b_1}), && \text{probabilities for } B(b_1) = \pm 1 \\
&g_\lambda(b_2 \pm, \lambda_{b_2}), && \text{probabilities for } B(b_2) = \pm 1 \\
&\cdots\cdots\cdots\cdots\cdots\cdots\cdots\cdots\cdots\cdots\cdots \\
&g_\lambda(b_n \pm, \lambda_{b_n}), && \text{probabilities for } B(b_n) = \pm 1
\end{aligned}
\tag{60}
$$

Of course the probabilities f_λ belong to the α-particles and the g_λ to the β-particles. Each of them depends only on the propensity of the measured observable. These propensities are now assumed to be strictly local. Thus, λ_{a_1} does not depend on the particular b_i that could be measured on the β-particles or even on whether a measurement is made at all on that side. This property of the propensities is a consequence of the homogenous nature of the sets $E_\lambda(z)$ once the locality assumption has been made.

It is important to stress that the present formulation of locality is strictly analogous to the quantum-mechanical locality of probabilities (which is known to hold despite the overall nonlocal nature of the theory). This formulation of locality is only applied to a wider set of probabilities.

An important feature of quantum probabilities is expected to hold for (59) and (60) as well: if $f(a_\mu\pm)$ is measured in a set S, this may render it impossible to measure $f(a_\nu\pm)$, with $\nu \neq \mu$, because the observables $A(a_\mu)$ and $A(a_\nu)$ are generally incompatible; similarly for $g(b_\sigma\pm)$ and $g(b_\tau\pm)$ in a set T if $\sigma \neq \tau$.

Obviously if one writes

$$\rho_\lambda(z) = \left(\frac{1}{N}\right)N_\lambda(z) \tag{61}$$

then

$$\sum_{\lambda \in I} \rho_\lambda(z) = 1$$

and $\rho_\lambda(z)$ is the "statistical weight" of the subset $E_\lambda(z)$ in the full set E.

3.3.4. New Factorizable Form of Joint Probabilities

The definition of Eq. (61) is useful for calculating all types of probabilities in the set E, for example, those needed for a new proof of Bell's theorem. These are typically joint probabilities for observations carried out on correlated systems. Considering the joint probabilities for $A(a_\mu) = \pm 1$ and $B(b_\nu) = +1$, one has

$$\omega(a_\mu\pm, b_\nu+) = \sum_{\lambda \in I} \rho_\lambda(z)\Omega_\lambda(a_\mu\pm, b_\nu+) \tag{62}$$

where the index λ indicates, as before, that the probability is calculated in the homogeneous set $E_\lambda(z)$. One can obviously apply Bayes' formula and write

$$\Omega_\lambda(a_\mu\pm, b_\nu+) = \omega_\lambda(a_\mu\pm, \lambda_{a_\mu}|b_\nu+)g_\lambda(b_\nu+, \lambda_{b_\nu}) \tag{63}$$

where $g_\lambda(b_\nu+, \lambda_{b_\nu})$ is as in (60) and $\omega_\lambda(a_\mu\pm, \lambda_{a_\mu}|b_\nu+)$ is the conditional probability in $E_\lambda(z)$ of $A(a_\mu) = \pm 1$, given that $B(b_\nu) = +1$. One can also say that $\omega_\lambda(a_\mu\pm, \lambda_{a_\mu}|b_\nu+)$ is the probability of $A(a_\mu) = \pm 1$ in

$$E_\lambda(z) \cap E(b_\nu+) \tag{64}$$

because it has been concluded that a set $E(b_\nu+)$ with the right properties exists if $B(b_\nu)$ is not measured. Remember now that $E_\lambda(z)$ is homogeneous, meaning that a probability valid for all of $E_\lambda(z)$ applies to any part of it as well, for example, to the intersection (64). Remembering Eq. (59), this gives

$$\omega_\lambda(a_\mu\pm, \lambda_{a_\mu}|b_\nu+) = f_\lambda(a_\mu\pm, \lambda_{a_\mu}) \tag{65}$$

Inserting Eq. (65) in Eq. (63), one gets

$$\Omega_\lambda(a_\mu\pm, b_\nu+) = f_\lambda(a_\mu\pm, \lambda_{a_\mu})g_\lambda(b_\nu+, \lambda_{b_\nu})$$

and consequently, from Eq. (62), it follows that

$$\omega(a_\mu\pm, b_\nu+) = \sum_{\lambda\in I(z)} \rho_\lambda(z)f_\lambda(a_\mu\pm, \lambda_{a_\mu})g_\lambda(b_\nu+, \lambda_{b_\nu}) \tag{66}$$

Because the argument is symmetrical in $+$ and $-$, we also have

$$\omega(a_\mu\pm, b_\nu-) = \sum_{\lambda\in I(z)} \rho_\lambda(z)f_\lambda(a_\mu\pm, \lambda_{a_\mu})g_\lambda(b_\nu-, \lambda_{b_\nu}) \tag{67}$$

The correlation function $P(a_\mu, b_\nu)$ is, by definition,

$$P(a_\mu, b_\nu) = \omega(a_\mu+, b_\nu+) - \omega(a_\mu+, b_\nu-) - \omega(a_\mu-, b_\nu+) + \omega(a_\mu-, b_\nu-)$$
$$= \sum_{\lambda\in I(z)} \rho_\lambda(z)p(a_\mu, \lambda)q(b_\nu, \lambda) \tag{68}$$

where

$$p(a_\mu, \lambda) = f_\lambda(a_\mu+, \lambda_{a_\mu}) - f_\lambda(a_\mu-, \lambda_{a_\mu})$$
$$q(b_\nu, \lambda) = g_\lambda(b_\nu+, \lambda_{b_\nu}) - g_\lambda(b_\nu-, \lambda_{b_\nu}) \tag{69}$$

In p and q the propensities have not been explicitly indicated.

The joint probabilities [Eqs. (66) and (67)] look like Clauser–Horne factorizability formulae,[10] with the index λ assuming the role of the hidden variable. There are, however, important differences because now the parameters of the observables in question enter the probability density $\rho_\lambda(z)$ [remember (58)]. This does not imply the introduction of any nonlocality because, in principle, all conceivable observables should enter $\rho_\lambda(z)$. In practice, however, one can take into account only the observables relevant to the set of correlation functions measured in the experiment considered. A similar situation has been described by Wigner[11] in his deterministically based proof of Bell's theorem, reviewed in the first chapter.

In spite of this dependence, it is very easy to deduce from Eqs. (66) and (67) the usual inequalities of Bell type. It is interesting to notice that it was never necessary to introduce joint probabilities for incompatible observables, described in quantum mechanics by noncommuting operators. Observables like $A(a_\mu)$ and $B(b_\nu)$ refer, instead, to two different objects and are therefore always compatible, so the previous obstacle does not exist and the joint probabilities [Eqs. (66) and (67)] are fully meaningful.

3.4. PROOFS OF WEAK AND STRONG INEQUALITIES

It has been shown[12] that the inequalities used for the experimental study of the Einstein–Podolsky–Rosen paradox are of two different types:

Weak inequalities, like Bell's original inequality, deduced directly from the principles of local realism. They would disagree with the empirical predictions of quantum theory only in the case of nearly perfect instruments, but are entirely compatible with such predictions for all performed experiments, owing to the low efficiency of photon detectors.

Strong inequalities, deduced with additional assumptions (in the sequel), are typically 20–30 times stronger than the weak ones, in the sense that they restrict *the same measurable quantity* to an interval 20–30 times smaller. The strong inequalities are violated by the predictions of quantum theory in all experiments that have been published.

The additional assumptions used by different authors are slightly different from case to case, but they always lead to exactly the same consequences. For example, the first assumption of this kind, made by Clauser, Horne, Shimony, and Holt[13] in 1969, was the following:

Given a pair of photons emerging from two regions of space where two polarizers can be located, the probability of their joint detection from two photomultipliers does not depend on the presence and the orientation of the polarizers.

It is important to stress that these additional assumptions are completely arbitrary and do not rest on any fundamental physical principle. Furthermore, there is no experiment to test them independently of local realism.

3.4.1. Weak Inequalities

To prove the weak Bell-type inequalities one can begin with the elementary Clauser–Horne lemma:[10] If

$$0 \leq x, x' \leq X \tag{70}$$

and

$$0 \leq y, y' \leq Y \tag{71}$$

then

$$-XY \leq \Gamma \leq 0 \tag{72}$$

where

$$\Gamma = xy - xy' + x'y + x'y' - x'Y - Xy \tag{73}$$

TABLE 3.1. The Quantity Γ [defined in (73)]
Calculated at the 16 Boundary Points

	x	x'	y	y'	Γ
1	0	0	0	0	0
2	0	0	0	Y	0
3	0	0	Y	0	$-XY$
4	0	0	Y	Y	$-XY$
5	0	X	0	0	$-XY$
6	0	X	0	Y	0
7	0	X	Y	0	$-XY$
8	0	X	Y	Y	0
9	X	0	0	0	0
10	X	0	0	Y	$-XY$
11	X	0	Y	0	0
12	X	0	Y	Y	$-XY$
13	X	X	0	0	$-XY$
14	X	X	0	Y	$-XY$
15	X	X	Y	0	0
16	X	X	Y	Y	0

Since Γ is linear in each of the variables x, x', y, y', its maximum and minimum will lie on the boundary, where these four variables take their extreme values. Looking at (70) and (71) one sees that the boundary consists of 16 points. The value of Γ at these points are given in Table 3.1. One sees that Γ assumes the values 0 and $-XY$ eight times each, and no other values, so inequality (72) must be satisfied.

The following notation will be used:

$E_\lambda \equiv E_\lambda(z)$ is the homogeneous set defined in (57).

$T_1(a, \lambda), T_1(a', \lambda)$ are the probabilities of the first photon, belonging to E_λ, being transmitted in polarizers with axes a and a', respectively.

$T_2(b, \lambda), T_2(b', \lambda)$ are the probabilities of the second photon, belonging to E_λ, being transmitted in polarizers with axes b and b', respectively.

$R_1(a, \lambda), R_1(a', \lambda)$ are the probabilities of the first photon, belonging to E_λ, being reflected in two-way polarizers with axes a and a', respectively.

$R_2(a, \lambda), R_2(a', \lambda)$ are the probabilities of the second photon, belonging to E_λ, being reflected in two-way polarizers with axes a and a', respectively.

$D_1^T(a, \lambda), D_1^T(a', \lambda)$ are the probabilties of the first photon, belonging to E_λ, being detected by a photomultiplier once it has been transmitted in polarizers with axes a and a', respectively. For one-way polarizers the superscript T will be dropped.

$D_2^T(b, \lambda)$, $D_2^T(b', \lambda)$ are the probabilities of the second photon, belonging to E_λ, being detected by a photomultiplier once it has been transmitted in polarizers with axes b and b', respectively. For one-way polarizers the superscript T will be dropped.

$D_1^R(a, \lambda)$, $D_1^R(a', \lambda)$ are the probabilities of the first photon, belonging to E_λ, being detected by a photomultiplier once it has been reflected in polarizers with axes a and a', respectively.

$D_2^R(b, \lambda)$, $D_2^R(b', \lambda)$ are the probabilities of the second photon, belonging to E_λ, being detected by a photomultiplier once it has been reflected in polarizers with axes b and b', respectively.

$D_1(\infty, \lambda)$ is the probability of the first photon, belonging to E_λ and on whose trajectory no polarizer is present being detected by the first photomultiplier.

$D_2(\infty, \lambda)$ is the probability of the second photon, belonging to E_λ and on whose trajectory no polarizer is present, being detected by the second photomultiplier.

See Fig. 3.5 for the setup of an ideal EPR experiment.
For the density function $\rho_\lambda(z)$,

$$\sum_{\lambda \in I(z)} \rho_\lambda(z) = 1$$

where $z = \{a, a', b, b'\}$. For one-way polarizers

$$R_1(a, \lambda) = R_2(a, \lambda) = D_1^R(a, \lambda) = D_2^R(a, \lambda) = 0$$

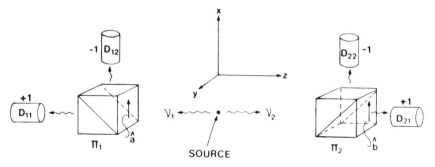

FIGURE 3.5. Setup of an ideal EPR experiment. The two-way polarizers π_1 and π_2 are set with their transmission axes in directions $\hat{a}$ and $\hat{b}$, respectively. Detectors D_{ij} ($i, j = 1, 2$) are 100% efficient. The source emits pairs of photons with frequencies ν_1 and ν_2 in the $-z$ and $+z$ directions, respectively.

which means that the reflected path is not available. The same applies to other axes.

The joint detection probabilities for two-way polarizers will be

$$\omega(a+, b+) = \sum_{\lambda \in I(z)} \rho_\lambda(z) T_1(a, \lambda) D_1^T(a, \lambda) T_2(b, \lambda) D_2^T(b, \lambda)$$

$$\omega(a+, b-) = \sum_{\lambda \in I(z)} \rho_\lambda(z) T_1(a, \lambda) D_1^T(a, \lambda) R_2(b, \lambda) D_2^R(b, \lambda)$$

$$\omega(a-, b+) = \sum_{\lambda \in I(z)} \rho_\lambda(z) R_1(a, \lambda) D_1^R(a, \lambda) T_2(b, \lambda) D_2^T(b, \lambda)$$

$$\omega(a-, b-) = \sum_{\lambda \in I(z)} \rho_\lambda(z) R_1(a, \lambda) D_1^R(a, \lambda) R_2(b, \lambda) D_2^R(b, \lambda)$$

$$(74)$$

There will be similar expressions for axes (a, b'), (a', b), (a', b'). Let us also define the correlation function

$$P(a, b) = \omega(a+, b+) - \omega(a+, b-) - \omega(a-, b+) + \omega(a-, b-) \qquad (75)$$

The single-particle detection probabilities are defined as follows:

$$t_1(a) = \sum_{\lambda \in I(z)} \rho_\lambda(z) T_1(a, \lambda) D_1^T(a, \lambda)$$

$$r_1(a) = \sum_{\lambda \in I(z)} \rho_\lambda(z) R_1(a, \lambda) D_1^R(a, \lambda)$$

$$t_2(b) = \sum_{\lambda \in I(z)} \rho_\lambda(z) T_2(b, \lambda) D_2^T(b, \lambda)$$

$$(76)$$

$$r_2(b) = \sum_{\lambda \in I(z)} \rho_\lambda(z) R_2(b, \lambda) D_2^R(b, \lambda)$$

where t and r refer to transmission and reflection. There will be similar expressions for axes a' and b'.

We can now deduce Bell-type inequalities by applying the Clauser–Horne lemma. Take, for instance,

$$T_1(a, \lambda) D_1^T(a, \lambda) = x \qquad T_2(b, \lambda) D_2^T(b, \lambda) = y$$
$$T_1(a', \lambda) D_1^T(a', \lambda) = x' \qquad T_2(b', \lambda) D_2^T(b', \lambda) = y'$$

If no additional assumption is made, these last variables could assume the value 1 for some λ. Therefore, one must take $X = Y = 1$ in (70) and (71), so that (72) becomes

$$-1 \le T_1(a, \lambda) D_1^T(a, \lambda) T_2(b, \lambda) D_2^T(b, \lambda) - T_1(a, \lambda) D_2^T(a, \lambda) T_2(b', \lambda) D_2^T(b', \lambda)$$
$$+ T_1(a', \lambda) D_1^T(a', \lambda) T_2(b, \lambda) D_2^T(b, \lambda) + T_1(a', \lambda) D_1^T(a', \lambda) T_2(b', \lambda) D_2^T(b', \lambda)$$
$$- T_1)(a', \lambda) d_1^t(a', \lambda) - T_2(b, \lambda) D_2^T(b, \lambda) \le 0$$

If we multiply the last equation by $\rho_\lambda(z)$ and sum over λ, we get the CHSH inequality:

$$-1 \leq \omega(a+, b+) - \omega(a+, b'+) + \omega(a'+, b+)$$
$$+ \omega(a'+, b'+) - t_1(a') - t_2(b) \leq 0 \tag{77}$$

We can repeat the argument by considering the transmission channel for the first photon and the reflection channel for the second photon and taking instead

$$T_1(a, \lambda)D_1^T(a, \lambda) = x \qquad R_2(b, \lambda)D_2^R(b, \lambda) = y$$
$$T_1(a', \lambda)D_1^T(a', \lambda) = x' \qquad R_2(b', \lambda)D_2^R(b', \lambda) = y'$$

again with $X = Y = 1$. We then obtain

$$-1 \leq \omega(a+, b-) - \omega(a+, b'-) + \omega(a'+, b-)$$
$$+ \omega(a'+, b'-) - t_1(a') - r_2(b) \leq 0 \tag{78}$$

A further repetition of the argument, considering the reflection channel for the first photon and the transmission channel for the second photon, leads to

$$-1 \leq \omega(a-, b+) - \omega(a-, b'+) + \omega(a'-, b+)$$
$$+ \omega(a'-, b'-) - r_1(a') - t_2(b) \leq 0 \tag{79}$$

A final repetition of the argument by considering the reflection channels for both photons leads to

$$-1 \leq \omega(a-, b-) - \omega(a-, b') + \omega(a'-, b-)$$
$$+ \omega(a'-, b'-) - r_1(a') - r_2(b) \leq 0 \tag{80}$$

Inequalities (77)–(80) are all necessary consequences of local realism. From them it is possible to deduce a Bell-type inequality for correlation functions: it is enough to change the signs of (78) and (79) and add the resulting new inequalities to (77) and (80) to obtain

$$-2 \leq P(ab) - P(ab') + P(a'b) + P(a'b') \leq 2 \tag{81}$$

where $P(ab)$, etc., were defined in (75). For one-way polarizers only (77) is meaningful and (78)–(81) do not correspond to actually measurable cases.

All these inequalities have been deduced from local realism only. They are called "weak" in order to distinguish them from the "strong" ones, which are dealt with in the following section.

3.4.2. Strong Inequalities for One-Way Polarizers

In experiments performed with one-way polarizers there is no photon reflection but only a detectable transmission or an undetectable absorption. We adapt our notation accordingly and drop the superscript T in the detection probabilities.

The quantum efficiencies, η_1 and η_2, of the photomultipliers will be taken to equal the λ-averages of D_1 and D_2:

$$\langle D_1(a, \lambda) \rangle_\lambda = \sum_{\lambda \in I(z)} \rho_\lambda(z) D_1(a, \lambda) = \eta_1$$

$$\langle D_2(b, \lambda) \rangle_\lambda = \sum_{\lambda \in I(z)} \rho_\lambda(z) D_2(b, \lambda) = \eta_2 \tag{82}$$

Similar relations, with the same η_1 and η_2, are assumed to hold for axes a' and b', respectively.

The measurable joint detection probabilities are given by

$$\omega_0 \equiv \omega(\infty, \infty) = \sum_{\lambda \in I(z)} \rho_\lambda(z) D_1(\infty, \lambda) D_2(\infty, \lambda)$$

$$\omega(a+, \infty) = \sum_{\lambda \in I(z)} \rho_\lambda(z) T_1(a, \lambda) D_1(a, \lambda) D_2(\infty, \lambda)$$

$$\omega(\infty, b+) = \sum_{\lambda \in I(z)} \rho_\lambda(z) D_1(\infty, \lambda) T_2(b, \lambda) D_2(b, \lambda) \tag{83}$$

$$\omega(a+, b+) = \sum_{\lambda \in I(z)} \rho_\lambda(z) T_1(a, \lambda) D_1(a, \lambda) T_2(b, \lambda) D_2(b, \lambda)$$

Similar definitions apply, in these last three cases, to the other pair of axes, (a, b'), (a', b) and (a', b').

1. **The additional assumption made by Clauser, Horne, Shimony, and Holt (CHSD).** Given that a pair of photons emerges from two regions of space where two polarizers can be located, the probability of their joint detection from two photomultipliers $D_{12}(\lambda)$ is independent of the presence and orientation of the polarizers.[13]

This additional assumption allows us to write

$$D_1(\infty, \lambda) D_2(\infty, \lambda) = D_1(a, \lambda) D_2(\infty, \lambda) = D_1(\infty, \lambda) D_2(b, \lambda)$$
$$= D_1(a, \lambda) D_2(b, \lambda) = D_{12}(\lambda) \tag{84}$$

and so on for a' and/or b'.

Write the inequalities (72) by taking, in Γ, $x = T_1(a, \lambda)$, $x' = T_1(a', \lambda)$, $y = T_2(b, \lambda)$, $y' = T_2(b', \lambda)$, with $X = Y = 1$:

$$-1 \leq T_1(a, \lambda) T_2(b, \lambda) - T_1(a, \lambda) T_2(b', \lambda)$$
$$+ T_1(a', \lambda) T_2(b, \lambda) + T_1(a', \lambda) T_2(b', \lambda) - T_1(a', \lambda) - T_2(b, \lambda)$$
$$\leq 0 \tag{85}$$

As a consequence of (84) one can multiply the whole expression by $\rho_\lambda(z)D_{12}(\lambda)$ and sum over λ to get

$$-\omega_0 \leq \omega(a+, b+) - \omega(a'+, b+) + \omega(a+, b'+)$$
$$+ \omega(a'+, b'+) - \omega(a', \infty) - \omega(\infty, b+) \leq 0 \qquad (86)$$

which is the fundamental strong inequality for one-way polarizers.

The difference with respect to the weak inequality (77) is the presence of $-\omega_0$ in the lhs. The nonparadoxical quantum prediction is $\omega_0 = \eta_1\eta_2$. In Table 3.2 one sees the values of η_1 and η_2 for published experiments. Notice that (77) and (86) can respectively be written

$$-1 + t_1(a') + t_2(b) \leq \Gamma \leq t_1(a') + t_2(b) \qquad (87)$$

$$-\omega_0 + \omega(a'+, \infty) + \omega(\infty, b+) \leq \Gamma \leq \omega(a'+, \infty) + \omega(\infty, b+) \qquad (88)$$

The single-particle probabilities in (87) have nonparadoxical expressions in quantum theory, as do the coincidence rates ω_0, $\omega(a'+, \infty)$, and $\omega(\infty, b+)$ in (88). The quantity

$$\Gamma = \omega(a+, b+) - \omega(a'+, b+) + \omega(a+, b'+) + \omega(a'+, b'+)$$

is the same in (87) and in (88). The interval to which Γ must be restricted has length 1 in (87) but only length $\omega_0 = \eta_1\eta_2 \leq 0.04$ in (88), as can be seen in Table 3.2.

2. **The additional assumption made by Clauser and Horne (CH)**. For every photon in the state λ the probability of detection with a polarizer placed on its trajectory is less than or equal to the detection probability with the polarizer removed.[10]

Therefore for the first photon,

$$D_1(a, \lambda), D_1(a', \lambda) \leq D_1(\infty, \lambda) \qquad (89)$$

Similarly for the second photon,

$$D_2(b, \lambda), D_2(b', \lambda) \leq D_2(\infty, \lambda) \qquad (90)$$

TABLE 3.2. Quantum Detector Efficiencies for Different EPR Experiments

	η_1	η_2	$\eta_1\eta_2$
Perrie *et al.*[39]	0.13	0.28	0.0364
Aspect *et al.*[35]	0.08	0.27	0.0216
Clauser[32]	0.26	0.07	0.0182
Holt and Pipkin[29]	0.06	0.25	0.0150
Freedman and Clauser[17]	0.20	0.20	0.0400

To deduce the strong inequality, one can use the Clauser–Horne lemma by taking

$$x = T_1(a, \lambda)D_1(a, \lambda) \qquad y = T_2(b, \lambda)D_2(b, \lambda)$$
$$x' = T_1(a', \lambda)D_1(a', \lambda) \qquad y' = T_2(b', \lambda)D_2(b', \lambda) \tag{91}$$

$$X = D_1(\infty, \lambda), \qquad Y = D_2(\infty, \lambda) \tag{92}$$

Obviously (91) and (92) satisfy (70) and (71). Indeed if (89) and (90) hold as they are, they will hold *a fortiori* if the lhs is multiplied by a transition probability. Therefore,

$$- D_1(\infty, \lambda)D_2(\infty, \lambda)$$
$$\leq T_1(a, \lambda)D_1(a, \lambda)T_2(b, \lambda)D_2(b, \lambda)$$
$$- T_1(a, \lambda)D_1(a, \lambda)T_2(b', \lambda)D_2(b', \lambda) + T_1(a', \lambda)D_1(a', \lambda)T_2(b, \lambda)D_2(b, \lambda)$$
$$+ T_1(a', \lambda)D_1(a', \lambda)T_2(b', \lambda)D_2(b', \lambda)$$
$$- T_1(a', \lambda)D_1(a', \lambda)D_2(\infty, \lambda) - D_1(\infty, \lambda)T_2(b, \lambda)D_2(b, \lambda)$$
$$\leq 0$$

Multiplying by $\rho_\lambda(z)$, summing over λ, and remembering the defintions (83), we again get the same inequality (86) obtained with the CHSH additional assumptions.

3. **Aspect's additional assumption.** The set of detected pairs with a given orientation of the polarizers is an undistorted representative sample of the set of pairs emitted by the source.[14]

The assumption is not very carefully formulated, because the possibility of absent polarizers should also have been considered. The main idea is, however, very clear and can be expressed thus: *Detectors introduce no distortions in the sets of pairs transmitted through regions where polarizers can be present.* Technically this means that every double-detection and -transmission probability is proportional to its corresponding double-transmission probability, and that the proportionality constant is always the same. In order to apply this assumption let us define the following double-transmission probabilities:

$$\tau(a+, b+) = \sum_{\lambda \in I(z)} \rho_\lambda(z)T_1(a, \lambda)T_2(b, \lambda)$$

$$\tau(a+, \infty) = \sum_{\lambda \in I(z)} \rho_\lambda(z)T_1(a, \lambda)$$

$$\tau(\infty, b+) = \sum_{\lambda \in I(z)} \rho_\lambda(z)T_2(b, \lambda)$$

$$\tau(\infty, \infty) = \sum_{\lambda \in I(z)} \rho_\lambda(z) = 1$$

Indeed it is natural to assume that

$$T_1(\infty, \lambda) = T_2(\infty, \lambda) = 1$$

TABLE 3.3. Numerical Comparison of Weak and Strong Inequalities for Different EPR Experiments

Inequality	Weak	Strong
Freedman and Clauser[17]	$-0.794 \leq \Gamma \leq 0.206$	$0.000 \leq \Gamma \leq 0.037$
Holt and Pipkin[29]	$-0.845 \leq \Gamma \leq 0.155$	$-0.002 \leq \Gamma \leq 0.019$
Clauser[32]	$-0.838 \leq \Gamma \leq 0.162$	$0.000 \leq \Gamma \leq 0.018$
Aspect et al.[35]	$-0.845 \leq \Gamma \leq 0.155$	$0.000 \leq \Gamma \leq 0.015$
Perrie et al.[39]	$-0.812 \leq \Gamma \leq 0.188$	$-0.002 \leq \Gamma \leq 0.038$

where the meaning of the symbols is as expected. One can start from (85), multiply by $\rho_\lambda(z)$, and sum over λ. The result is

$$-\tau_0 \leq \tau(a+, b+) - \tau(a+, b'+) + \tau(a'+, b+)$$
$$+ \tau(a'+, b'+) - \tau(a'+, \infty) - \tau(\infty, b+) \leq 0 \qquad (93)$$

where $\tau_0 = \tau(\infty, \infty)$. According to quantum mechanics the double-detection probability when no polarizers are placed on the paths of the two photons is $\omega_0 = \eta_1 \eta_2$. This allows us to identify $\eta_1 \eta_2$ as the universal constant, which, according to Aspect's assumption, gives the ratio between a double-transmission and -detection probability, and the corresponding double-transmission probability. We can write

$$\omega(a+, b+) = \tau(a+, b+)\eta_1 \eta_2$$
$$\omega(a+, \infty) = \tau(a+, \infty)\eta_1 \eta_2$$
$$\omega(\infty, b+) = \tau(\infty, b+)\eta_1 \eta_2$$
$$\omega_0 = \tau_0 \eta_1 \eta_2 = \eta_1 \eta_2$$

and so on for the other similar probabilities.

Therefore by multiplying (93) by $\eta_1 \eta_2$, one again obtains the inequalities (86). The striking difference between strong and weak inequalities for published experiments is shown in Table 3.3 and Fig. 3.6. Again, Γ is the quantity appearing in (87) and (88).

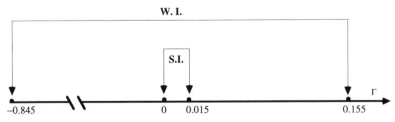

FIGURE 3.6. The measurable quantity Γ is predicted to fall within the intervals $(-0.845, 0.155)$ and $(0, 0.015)$ by the weak (W. I.) and strong inequality (S. I.), respectively. The given figures are exact for the experiment by Aspect et al.[28]

3.4.3. Strong Inequalities for Two-Way Polarizers

The theory of EPR experiments with two-way polarizers was developed by Garuccio and Rapisarda[15] (GR) and adopted by Aspect et al. in their second experiment.[16] To obtain a strong inequality, GR made the following assumption:

GR Assumption. The sum of the probability of transmission and detection, and of the probability of reflection and detection in a two-way polarizer does not depend on the orientation of the polarizer.

More formally,

$$T_1(a, \lambda)D_1^T(a, \lambda) + R_1(a, \lambda)T_1^R(a, \lambda) = F(\lambda)$$
$$T_2(b, \lambda)D_2^T(b, \lambda) + R_2(b, \lambda)T_2^R(b, \lambda) = G(\lambda) \tag{94}$$

Garuccio and Rapisarda introduced the following definition of a normalized correlation function:

$$E(a, b) = \frac{N(a, b)}{Z(a, b)} \tag{95}$$

where

$$N(a, b) = \omega(a+, b+) - \omega(a+, b-) - \omega(a-, b+) + \omega(a-, b-)$$
$$Z(a, b) = \omega(a+, b+) + \omega(a+, b-) + \omega(a-, b+) + \omega(a-, b-) \tag{96}$$

and the joint probabilities $\omega(a\pm, b\pm)$ were defined in (74). It is a simple matter to see that (74) and (96) imply

$$N(a, b) = \sum_{\lambda \in I(z)} \rho_\lambda(z)[T_1(a, \lambda)D_1^T(a, \lambda) - R_1(a, \lambda)D_1^R(a, \lambda)]$$
$$\times [T_2(b, \lambda)D_2^T(b, \lambda) - R_2(b, \lambda)D_2^R(b, \lambda)]$$
$$Z(a, b) = \sum_{\lambda \in I(z)} \rho_\lambda(z)[T_1(a, \lambda)D_1^T(a, \lambda) + R_1(a, \lambda)D_1^R(a, \lambda)]$$
$$\times [T_2(b, \lambda)D_2^T(b, \lambda) + R_2(b, \lambda)D_2^R(b, \lambda)] \tag{97}$$

As a consequence of (94) one can write Z without reference to a and b:

$$Z = \sum_{\lambda \in I(z)} \rho_\lambda(z)F(\lambda)G(\lambda) \tag{98}$$

Writing

$$T_1(a, \lambda)D_1^T(a, \lambda) - R_1(a, \lambda)D_1^R(a, \lambda) \equiv p(a, \lambda)$$
$$T_2(b, \lambda)D_2^T(b, \lambda) - R_2(b, \lambda)D_2^R(b, \lambda) \equiv q(b, \lambda) \tag{99}$$

it becomes apparent that (94), allows one to write

$$|p(a, \lambda)| \le F(\lambda); \qquad |q(b, \lambda)| \le G(\lambda) \qquad (100)$$

But (95)–(97) give

$$E(a, b) = \frac{1}{Z} \sum_{\lambda \in I(z)} \rho_\lambda(z) p(a, \lambda) q(b, \lambda) \qquad (101)$$

so that

$$|E(a, b) - E(a, b') + E(a', b) + E(a', b')|$$

$$\le \frac{1}{Z} \sum_{\lambda \in I(z)} \rho_\lambda(z) \{ |p(a, \lambda)| \, |q(b, \lambda) - q(b', \lambda)| + |p(a', \lambda)| \, |q(b, \lambda) + q(b', \lambda)| \}$$

$$(102)$$

Using the directions a, a', b, b' in (100), (102) becomes

$$|E(a, b) - E(a, b') + E(a', b) + E(a', b')| \le 2 \qquad (103)$$

Written in terms of correlations functions $P(a, b)$ as defined in (75), this new strong inequality becomes

$$|P(a, b) - P(a, b') + P(a', b) + P(a', b')| \le 2Z$$

This inequality is of direct physical interest because its rhs can be identified with $2\eta_1 \eta_2$, twice the product of the quantum efficiencies of the detectors placed on the two channels.

3.4.4. Weak and Strong Freedman Inequalities

The weak and strong inequalities obtained for one-way polarizers, are, respectively, (77) and (86). From them it is possible to deduce Freedman's form[17] of these inequalities. To do so, make the assumption, easy to check experimentally and predicted by quantum mechanics, that the experimental settings a and b enter the double-detection probability only through the angle between the polarizer axes:

$$\omega(a+, b+) = \omega(a - b) \qquad (104)$$

Now by using (86) and (104), one obtains

$$-\omega_0 + \omega(a', \infty) + \omega(\infty, b) \le \omega(a - b) - \omega(a - b')$$
$$+\omega(a' - b) + \omega(a' - b') \le \omega(a', \infty) + \omega(\infty, b) \qquad (105)$$

where the + symbols have been dropped. Choosing for the experimental parameters values satisfying

$$(a - b) = (a' - b) = (a' - b') = \varphi \qquad \text{and} \qquad (a - b') = 3\varphi \qquad (106)$$

it follows that

$$-\omega_0 + \omega(a', \infty) + \omega(\infty, b) \leq 3\omega(\varphi) - \omega(3\varphi) \leq \omega(a', \infty) + \omega(\infty, b) \quad (107)$$

Maximum violations of quantum mechanics take place for $\varphi = 22.5°$ and $\varphi = 67.5°$. The inequalities (106) for these two values of φ can be combined as

$$|\omega(22.5°) - \omega(67.5°)| \leq \omega_0/4 \quad (108)$$

Repeating the same steps for inequality (77), one obtains the weak form of Freedman's inequality:

$$|\omega(22.5°) - \omega(67.5°)| \leq 1/4 \quad (109)$$

We shall now compare Freedman's inequalities with the quantum-mechanical predictions. On setting

$$\delta = |\omega(22.5°)/\omega_0 - \omega(67.5°)/\omega_0| \quad (110)$$

the strong form of Freedman's inequality becomes

$$\delta \leq \tfrac{1}{4} \quad (111)$$

and the weak form becomes

$$\delta \leq 1/4\omega_0 \quad (112)$$

Most of the experiments have made use of atomic-cascade photon pairs in a singlet state. The quantum-mechanical double-detection probability for this kind of source is

$$\omega_{QM}(a - b) = \frac{1}{4}[\epsilon_+^1 \epsilon_+^2 + \epsilon_-^1 \epsilon_-^2 F(\theta) \cos 2(a - b)]\eta_1 \eta_2 \quad (113)$$

where $\epsilon_\pm^1$ and $\epsilon_\pm^2$ are parameters related to the transmittance of the polarizers, $F(\theta)$ is a function of the half-angle θ of the cones subtended by each detector aperture, and η_1 and η_2 are the quantum efficiencies of the photomultipliers. If the polarizers have been removed

$$\omega_0 = \eta_1 \eta_2 \quad (114)$$

The quantum-mechanical value of δ, therefore, is

$$\delta_{QM} = \frac{\sqrt{2}}{4} \epsilon_-^1 \epsilon_-^2 F(\theta) \quad (115)$$

In Table 3.3 the weak and strong limits of Bell's inequalities are compared with the quantum-mechanical values. It can be seen that those predictions always exceed the strong limit but are largely below the weak one; therefore, none of the experiments reported in Table 3.3, in principle, can disprove Freedman's weak inequality. Moreover, local realistic models in which the additional assumptions are explicitly violated have been shown to reproduce the experimental predictions of quantum mechanics when low-efficiency detectors are employed.[18]

The additional assumptions have played a fundamental role in shifting the debate on local realism from the theoretical positions to the experimental ground; they have been a tool to obtain stronger limits when low-efficiency photomultipliers have been employed. The confidence placed in these assumptions is far weaker than that in local realism, so it seems more reasonable to discard them rather than consider the performed experiments as a conclusive proof of the violation of local realism.

3.4.5. Generalized Bell-Type Inequalities

In the late seventies it became clear that other weak inequalities like Bell's could be deduced from local realism. In practice, an inequality of Bell type could be deduced for every linear combination of correlation functions. Sometimes these provide further physical restrictions.[19-21] For example, Roy and Singh[20] deduced three interesting inequalities. Their results will be mentioned, but their proof will not be discussed because later in this section a very general method of proof of inequalities for all possible linear combinations of correlation functions will be presented, those of Roy and Singh included. Before considering them let us note that given a linear combination of correlation functions

$$\sum_{i=1}^{m}\sum_{j=1}^{n} C_{ij}P(a_i, b_j)$$

the real coefficients C_{ij} define an $m \times n$ matrix that can be taken to represent completely the original linear combination.

The first inequality is

$$\sum_{i=1}^{4}\sum_{j=1}^{5} C_{ij}^1 P(a_i, b_j) \leq 6 \tag{116}$$

where

$$C_{ij}^1 = \begin{pmatrix} 1 & 1 & 1 & 0 & 1 \\ 1 & 1 & -1 & 1 & 0 \\ 1 & 1 & 0 & -1 & -1 \\ 1 & -1 & 0 & 0 & 0 \end{pmatrix}$$

The second one is

$$\sum_{i=1}^{4}\sum_{j=1}^{7} C_{ij}^2 P(a_i, b_j) \leq 8 \tag{117}$$

where

$$C_{ij}^2 = \begin{pmatrix} 1 & 1 & 1 & 1 & 0 & 0 & 0 \\ 1 & -1 & 0 & 0 & 1 & 1 & 0 \\ 1 & 0 & -1 & 0 & -1 & 0 & 1 \\ 1 & 0 & 0 & -1 & 0 & -1 & -1 \end{pmatrix}$$

and the third one is

$$\sum_{i=1}^{6} \sum_{j=1}^{8} C_{ij}^3 P(a_i, b_j) \leq 16 \tag{118}$$

where

$$C_{ij}^3 = \begin{pmatrix} 1 & 1 & -1 & -1 & 1 & 1 & 1 & 1 \\ 1 & 1 & 1 & 1 & -1 & -1 & 1 & 1 \\ 1 & 1 & 1 & 1 & 1 & 1 & -1 & -1 \\ 1 & -1 & -1 & 1 & 1 & -1 & 1 & -1 \\ 1 & -1 & 1 & -1 & -1 & 1 & 1 & -1 \\ 1 & -1 & 1 & -1 & 1 & -1 & -1 & 1 \end{pmatrix}$$

It is easy to show that inequalities (116)–(118) provide restrictions on $P(a_i, b_j)$ not implied by Bell's inequality. Suppose, for example, that the correlation functions multiplied by negative coefficients in (116) vanish—

$$P(a_4, b_2) = P(a_2, b_3) = P(a_3, b_4) = P(a_3, b_5) = 0$$

and the remaining $P(a_i, b_j)$ occurring in equality (116) are all equal to 2/3; then all the Bell inequalities involving these $P(a_i, b_j)$ are obviously satisfied, but inequality (116) is violated because its lhs is $10 \cdot \frac{2}{3} > 6$.

Next a general proof will be given of a large set of inequalities using the formulation of probabilistic local realism of Section 3.3. If $\rho_\lambda(z) \geq 0$ is the statistical weight of the λth homogeneous subset E_λ, the correlation function is given by (68):

$$P(a, b) = \sum_{\lambda \in I(z)} \rho_\lambda(z) p(a, \lambda) q(b, \lambda) \tag{119}$$

where, using the notation of (69),

$$\begin{aligned} p(a, \lambda) &= f_\lambda(a+, \lambda_a) - f_\lambda(a-, \lambda_a) \\ q(b, \lambda) &= g_\lambda(b+, \lambda_b) - g_\lambda(b-, \lambda_b) \end{aligned} \tag{120}$$

Since $f_\lambda(a+, \lambda_a) + f_\lambda(a-, \lambda_a) = 1$ and $g_\lambda(b+, \lambda_b) + g_\lambda(b-, \lambda_b) = 1$, it follows that

$$|p(a, \lambda)| \leq 1, \qquad |q(b, \lambda)| \leq 1 \tag{121}$$

The only interesting inequalities deduced from local realism are those that hold for all conceivable local realistic theories of the type given by Eqs. (119) and (120). Obviously, it is not possible today to say which (if any) of the infinitely many theories based on local realism is the right one. Therefore, inequalities deduced from a particular theory (or from a particular set of theories) are not interesting.

The following lemma allows one to deduce inequalities that are true for all conceivable local realistic theories.

Lemma. *Given a real number M, the inequality*

$$\sum_{ij} C_{ij} P(a_i, b_j) \leq M \tag{122}$$

holds for all conceivable local realistic theories if and only if the inequality

$$\sum_{ij} C_{ij} p(a_i, \lambda) q(b_j, \lambda) \leq M \tag{123}$$

holds for arbitrary values of λ and for arbitrary dependence of p and q on their arguments.

Proof. Inequality (123) is a consequence of inequality (122) since, among all conceivable local realistic theories, there are those in which the statistical weight $\rho_\lambda(z)$ in (119) is a Kronecker delta $\delta_{\lambda\lambda_0}$ and therefore

$$\sum_{ij} C_{ij} \sum_{\lambda} \rho_\lambda(z) p(a_i, \lambda) q(b_j, \lambda) \leq M$$

implies that

$$\sum_{ij} C_{ij} p(a_i, \lambda_0) q(b_j, \lambda_0) \leq M \tag{124}$$

where λ_0, being arbitrary, can assume any value. Conversely, if inequality (123) holds for arbitrary λ and arbitrary dependence of $p(a_i, \lambda)$ and $q(a_i, \lambda)$ on their arguments, it is sufficient to multiply it by $\rho_\lambda(z)$ and sum in order to obtain inequality (122) as true for an arbitrary local realistic theory. The proof is thus completed.

This lemma does not specify the value of M. One calls an inequality of type (122) "trivial" if

$$M \geq \sum_{ij} |C_{ij}|$$

In fact, the lhs of inequality (122) cannot be larger than the rhs of the previous inequality since every correlation function $P(a_i, b_j)$ has, by definition, a modulus not exceeding 1.

The objective here is therefore *the determination of nontrivial inequalities satisfied by all conceivable local hidden-variable theories.* Obviously, the most stringent inequality is found when M is taken equal to the maximum value of the lhs of inequality (123) (for given C_{ij}):

$$M = \max\left\{ \sum_{ij} C_{ij} p(a_i, \lambda) q(b_j, \lambda) \right\} \tag{125}$$

The quantity in braces in (125) is linear in $p(a_i, \lambda)$ and $q(b_j, \lambda)$; therefore its maximum M is found on the boundary, namely, at one of the vertices of the hypercube C in the multidimensional space having $p(a_i)$ and $q(b_j)$ as Cartesian coordinates; i.e.,

$$M = \max_{\xi_i \eta_j}\left\{ \sum_{ij} C_{ij} \xi_i \eta_j \right\} \tag{126}$$

where $\xi_i = \pm 1$, $\eta_j = \pm 1$, and the maximum is taken over all possible choices of ξ_i and η_j. Thus the most general consequence of local realism for linear combinations of correlation function reads

$$\sum_{i=1}^{m} \sum_{j=1}^{n} C_{ij} P(a_i, b_j) \leq \max_{\xi_i \eta_j}\left\{ \sum_{i=1}^{m} \sum_{j=1}^{n} C_{ij} \xi_i \eta_j \right\} \tag{127}$$

Three theorems allow us to narrow the set of inequalities of the type (127) which can be considered to be of physical interest.

Theorem 1. *Every inequality whose coefficients C_{ij} have factorizable signs is trivial.*

In fact if

$$C_{ij} = |C_{ij}| \mu_i v_j, \qquad \mu_i, v_j = \pm 1$$

one has from Eq. (126) that

$$M = \max_{\xi_i \eta_j}\left\{ \sum_{ij} |C_{ij}| \mu_i v_j \xi_i \eta_j \right\} = \sum_{ij} |C_{ij}|$$

since it is possible to choose $\xi_i = \mu_i$ and $\eta_j = v_j$ for all i and j.

Theorem 2. *If an argument a_i or an argument b_j appears only once, the inequality can be reduced to a more elementary one.*

Indeed, there is a one-to-one correspondence between the experimental parameters a_i and sign factors ξ_i, and between b_j and η_j. As a consequence, if we suppose that a_1 figures only once, then the sign factor ξ_1 figures only once, and

$$M = \max_{\xi_i, \eta_j}\left\{ C_{1l} \xi_1 \eta_l + \sum_{i=2}^{m} \sum_{j=1}^{n} C_{ij} \xi_i \eta_j \right\}$$

$$= |C_{1l}| + \max_{\xi_i \eta_j}\left\{ \sum_{i=2}^{m} \sum_{j=1}^{n} C_{ij} \xi_i \eta_j \right\}$$

since one can always choose ξ_1 in such a way that $C_{1l}\xi_1\eta_l = |C_{1l}|$. In this case the inequality

$$\sum_{i=1}^{m}\sum_{j=1}^{n} C_{ij}P(a_i, b_j) \leq M$$

can be reduced to the more elementary one

$$\sum_{i=2}^{m}\sum_{j=1}^{n} C_{ij}P(a_i, b_j) \leq M - |C_{1l}|$$

Theorem 3. *If the lhs of inequality (122) can be split into two parts such that no argument a_i or b_j is common to two correlation functions belonging to each of these two parts, then the inequality deducible from local realism can be reduced to two more elementary inequalities.*

Proof. The correspondence between parameters a_i and b_j, and sign factors ξ_i and η_j, ensures that if the lhs can be split into two parts, with no argument a_i or b_j in common, then the rhs can also be split into two parts having no sign factor ξ_i or η_j in common. Hence, the original inequality can be written as the sum of the two simpler ones.

It is easy to prove, using Theorem 2, that linear combinations of three correlation functions give only reducible inequalities. More generally, odd numbers of correlation functions give only reducible inequalities.

Consider, then, the case of four correlation functions. By Eq. (126) the maximum for Einstein locality is obtained for

$$\sum_{i,j=1}^{2} C_{ij}P(a_i, b_j) \leq M = \max\left\{\sum_{i=1}^{2}\sum_{j=1}^{2} C_{ij}\xi_i\eta_j\right\}$$

$$= \max\left\{\sum_{i=1}^{2}\xi_i\sum_{j=1}^{2}|C_{ij}|\sigma_{ij}\eta_j\right\}$$

$$= \max\left\{\sum_{i=1}^{2}\left|\sum_{j=1}^{2}|C_{ij}|\sigma_{ij}\eta_j\right|\right\}$$

$$= \max\{||C_{11}| + |C_{12}|\xi| + ||C_{22}| + |C_{21}|\sigma\xi|\} \qquad (128)$$

where σ_{ij} is the sign of C_{ij}, $\xi = \sigma_{11}\cdot\sigma_{12}\cdot\eta_1\cdot\eta_2$, and $\sigma = \sigma_{11}\cdot\sigma_{12}\cdot\sigma_{21}\cdot\sigma_{22}$.

If $\sigma = +1$ (or if, equivalently, the signs are factorizable) the result is trivial:

$$M = \sum_{ij}|C_{ij}|$$

If, instead, $\sigma = -1$, then

$$M = \sum_{ij}|C_{ij}| - 2\min_{lm}|C_{lm}| \qquad (129)$$

From (129) it is possible to deduce that, in the case of four correlation functions,

$$M \geq \frac{1}{2} \sum_{ij} |C_{ij}|$$

where equality holds only when C_{ij} is constant for all i and j, or, equivalently, for Bell's inequality; so in the case of four correlation functions no inequality stronger than Bell's inequality (for which equality obviously holds) exists.

The case $n = 6$ is more complicated and only the result will be presented here. One obtains

$$\sum_{i=1}^{2} \sum_{j=1}^{3} C_{ij} P(a_i, b_j) \leq M \tag{130}$$

where

$$M = \max_{\rho_1 \rho_2} \{| |C_{11}| + |C_{12}|\rho_1 + |C_{13}|\rho_2|$$

$$+ ||C_{21}| + |C_{22}|\rho_1 \sigma_1 + |C_{23}|\rho_2 \sigma_2|\} \tag{131}$$

where ρ_1 and ρ_2 are sign factors chosen in such a way as to maximize M and

$$\sigma_1 = \sigma_{11} \sigma_{12} \sigma_{21} \sigma_{22}, \qquad \sigma_2 = \sigma_{11} \sigma_{13} \sigma_{21} \sigma_{23}$$

A particular application of inequality (130)

$$P(a_1, b_1) + P(a_1, b_2) - P(a_1, b_3) + P(a_2, b_1) + P(a_2, b_2) - P(a_2, b_3) \leq 2 \tag{132}$$

as can easily be checked. This inequality can be more restrictive than any Bell inequality because if the correlation functions appearing with the $+ (-)$ sign in (132) assume the value $+\frac{1}{2}$ $(-\frac{1}{2})$, the latter inequality is violated while Bell's is satisfied.

Given any general inequality, an "associated Bell inequality" is a Bell inequality that contains correlation functions that also appear in the original inequality.

The following theorem gives a powerful method for singling out the inequalities that provide restrictions on correlation functions not implied by Bell's inequality.

Theorem 4. *Given a linear combination*

$$L = \sum_{ij} C_{ij} P(a_i, b_j)$$

if

$$\bar{M} = \sum_{ij} |C_{ij}|$$

and if

$$M = \max_{\xi_i \eta_j} \left\{ \sum_{i=1}^{m} \sum_{j=1}^{n} C_{ij} \xi_i \eta_j \right\}$$

is the maximum value of L allowed by local realism, then the inequality $L \le M$ implies the existence of physical restrictions not contained in any Bell inequality provided that

$$M < \tfrac{1}{2}\bar{M} \tag{133}$$

Proof. Consider the $n \times m$ space in which the $P(a_i, b_j)$ are located on the axes and the vector $\bar{P} = \{\bar{P}(a_i, b_j)\}$, which maximizes the linear combination L for given C_{ij}. The components of this vector all have modulus 1 and their signs are the same as those of the corresponding C_{ij}. The components of the new vector

$$\{\bar{P}'(a_i, b_j)\} = \{0.5\bar{P}(a_i, b_j)\}$$

all satisfy the associated Bell inequalities [because each $|\bar{P}'(a_i, b_j)| = 0.5$], but it results in

$$\sum_{i=1}^{n} \sum_{j=1}^{m} C_{ij} \bar{P}'(a_i, b_j) = \tfrac{1}{2}\bar{M} > M$$

The inequalities of local realism that satisfy condition (133) are called "superinequalities." The Roy–Singh inequalities (116) and (118) are historically the first examples of superinequalities; inequality (132) is another.

Theorem 4 only expresses a *sufficient* condition for the existence of an inequality stronger than Bell's inequality. For example, inequality (117) has its rhs equal to one half of the possible maximum, but provides restrictions on $P(a_i, b_j)$ not implied by Bell's inequality.

The case of three values for a and three (or more) for b was analyzed by Garg.[22] An interesting result is obtained in the case of four different values each for a_i and b_j. In this case the number of correlation functions is 16 and the number of different 4×4 matrices with integer coefficients C_{ij} in the range $\{-2, 2\}$ is 5^{16}. Using the previous method and a computer, Garuccio[23] analyzed 13,500,000 matrices (0.009% of the total) and found 1050 superinequalities. Since the region analyzed has no special features, it is probably possible to generalize the result and conclude that an analysis of the complete set of 4×4 inequalities of the stated kind would give nearly 10^7 superinequalities.

The following are some examples of these inequalities:

1.
$$\sum_{i,j=1}^{4} C_{ij}^{1} P(a_i, b_j) \le 11 \tag{134}$$

where

$$C_{ij}^1 = \begin{pmatrix} 2 & 2 & 2 & 2 \\ 2 & 2 & 1 & -2 \\ -2 & 2 & 0 & 0 \\ 1 & 1 & -2 & 0 \end{pmatrix}$$

2. $$\sum_{i,j=1}^{4} C_{ij}^2 P(a_i, b_j) \le 10 \qquad (135)$$

where

$$C_{ij}^2 = \begin{pmatrix} 2 & 2 & 2 & 0 \\ 2 & 1 & -1 & -2 \\ -2 & 1 & -1 & 2 \\ 2 & -2 & 0 & 0 \end{pmatrix}$$

3. $$\sum_{i,j=1}^{4} C_{ij}^3 P(a_i, b_j) \le 6 \qquad (136)$$

where

$$C_{ij}^3 = \begin{pmatrix} 1 & 1 & 1 & -1 \\ -1 & 1 & 1 & -1 \\ 0 & 2 & -1 & 1 \\ 0 & 0 & 1 & 1 \end{pmatrix}$$

To clarify the content of Theorem 4, consider inequality (134). The maximum possible value of the lhs is obviously

$$\bar{M} = \sum_{i,j}^{4} |C_{ij}^1| = 23$$

and is obtained for a suitable choice of $P(a_i, b_j)$:

$$\{\bar{P}(a_i, b_j)\} = \begin{pmatrix} 1 & 1 & 1 & 1 \\ 1 & 1 & 1 & -1 \\ -1 & 1 & 0 & 0 \\ 1 & 1 & -1 & 0 \end{pmatrix} \qquad (137)$$

[A matrix representation is used for the vector $\{\bar{P}(a_i, b_j)\}$ in the 4×4 space.]

Starting from set (137) it is possible to define the new set of correlation functions:

$$\{\bar{P}'(a_i, b_j)\} = \begin{pmatrix} 0.5 & 0.5 & 0.5 & 0.5 \\ 0.5 & 0.5 & 0.5 & -0.5 \\ -0.5 & 0.5 & 0 & 0 \\ 0.5 & 0.5 & -0.5 & 0 \end{pmatrix} \tag{138}$$

Since all nonvanishing functions $\bar{P}'(a_i, b_j)$ are equal to ± 0.5, all Bell inequalities containing the $\bar{P}'(a_i, b_j)$ are satisfied. Therefore, using only the Bell inequality, one could tentatively conclude that the set (138) describes a physical system compatible with local realism. This is, however, not true since the set (138) introduced in inequality (134) gives $11.5 < 11$; and therefore the inequality is violated and the set of correlation functions (138) cannot be obtained from a local theory. Moreover, a hypersphere with center $\{\bar{P}'(a_i, b_j)\}$ and radius $R = \frac{1}{2}(0.5)$ exists such that all the sets of correlation functions inside this circle that satisfy Bell's inequality violate inequality (134).

It was stated that, in the case of 4×4 correlation functions, a large number of superinequalities exists. It is possible that only a finite number of these inequalities are independent and form a set that completely expresses local realism. Further studies would answer this question.

New and more stringent inequalities were deduced in 1987 by Lepore[24] for linear combinations *of joint probabilities*. The physical content of these inequalities is not entirely deducible from the inequalities discussed above.

Consider m instrumental parameters $a_1, a_2, \ldots, a_m$ for the first measuring apparatus and n instrumental parameters $b_1, b_2, \ldots, b_n$ for the second measuring apparatus. Let

$$\omega_{hk}(a_i, b_j) = \sum_{\lambda \in I(z)} \rho_\lambda(z) p_h(a_i, \lambda) q_k(b_j, \lambda) \tag{139}$$

by the joint probability of measuring $A(a_i)$ and obtaining h and measuring $B(b_j)$ and obtaining $k(h, k = \pm 1)$. Now consider the linear combination of joint probabilities

$$C = \sum_{hkij} C_{ij}^{hk} \omega_{hk}(a_i, b_j) \tag{140}$$

where C_{ij}^{hk} are $4mn$ arbitrary coefficients.

In order to deduce the inequalities

$$m_0 \leq C \leq M_0 \tag{141}$$

true for all conceivable local realistic theories, it is sufficient, using the lemma, to prove that the inequality

$$m_0 \leq \sum_{hkij} C_{ij}^{hk} p_h(a_i, \lambda) q_k(b_j, \lambda) \leq M_0 \tag{142}$$

is true for arbitrary value of λ and for arbitrary dependence of p_h and q_k on their arguments.

Using the relations

$$p_+(a_i, \lambda) + p_-(a_i, \lambda) = 1$$
$$q_+(b_j, \lambda) + q_-(b_j, \lambda) = 1$$

one can write

$$\sum_{hkij} C_{ij}^{hk} p_h(a_i, h) q_k(b_j, \lambda)$$
$$= F[p_+(a_1, \lambda) \cdots p_+(a_m, \lambda), q_+(b_1, \lambda) \cdots q_+(b_n, \lambda)] \tag{143}$$

where F is a linear function of each of the indicated arguments. The most stringent inequality is found when M_0 is taken equal to the maximum value (and m_0 equal to the minimum value) of F. Sinxe F is defined in the hypercube in the multidimensional space having $p_+(a_i, \lambda)$ and $q_+(b_j, \lambda)$ as Cartesian coordinates, the maximum and minimum are found on the boundary, namely at one of the vertices of hypercube.

Therefore, setting

$$m_0 = \min_{\substack{\xi_1, \ldots, \xi_m = 0,1 \\ \eta_1, \ldots, \eta_n = 0,1}} F(\xi_1, \ldots, \xi_m, \eta_1, \ldots, \eta_n) \tag{144}$$

$$M_0 = \max_{\substack{\xi_1, \ldots, \xi_m = 0,1 \\ \eta_1, \ldots, \eta_n = 0,1}} F(\xi_1, \ldots, \xi_m, \eta_1, \ldots, \eta_n) \tag{145}$$

one obtains the set of inequalities

$$m_0 \leq \sum_{hkij} C_{ij}^{hk} \omega_{hk}(a_i, b_j) \leq M_0 \tag{146}$$

For every choice of coefficients C_{ij}^{hk} relation (146) provides the most stringent inequalities that can be deduced from local realism if m_0 and M_0 are given by (144) and (145), respectively.

To prove that the set of inequalities (146) is not equivalent to the set of inequalities of correlation functions (127) and implies more stringent restriction for local realism, the following model studied by Garg and Mermin[25] in 1982 will be used. In this model the joint probabilities are

$$\omega_{hk}(a_i, b_j) = \tfrac{1}{4}[1 - c(h + k) + A_{ij} hk] \tag{147}$$

$(i, j = 1, 2, 3)$ with

$$0 \leq c \leq \tfrac{1}{3} \tag{148}$$

and

$$A_{11} = A_{22} = 1$$
$$A_{ij} = -\tfrac{1}{3} \quad \text{for } i = j = 3 \text{ and } i \neq j \tag{149}$$

From Eq. (147) one has for the correlation functions

$$P(a_i, b_j) = A_{ij}$$

and it is easy to prove that this model satisfies all inequalities deduced from local realism for correlation functions.

Consider now the case of three directions a_1, a_2, a_3 for the first apparatus and three directions b_1, b_2, b_3 for the second. One can write the linear combination

$$\omega(a_3+, b_3+) - \omega(a_2-, b_2-) + \omega(a_1-, b_1+) + \omega(a_2+, b_2-)$$
$$+ \omega(a_3-, b_2-) - \omega(a_3-, b_1+) + 2\omega(a_1+, b_2+)$$
$$+ \omega(a_2-, b_3-) - \omega(a_1+, b_3-)$$

By calculating the minimum value of the associated function F, one gets min $F = 0$; therefore the following inequality holds:

$$\omega(a_3+, b_3+) - \omega(a_2-, b_2-) + \omega(a_1-, b_1+) + \omega(a_2+, b_2-) + \omega(a_3-, b_2-)$$
$$- \omega(a_3-, b_1+) + 2\omega(a_1+, b_2+) + \omega(a_2-, b_3-) - \omega(a_1+, b_3-)$$
$$\geq 0 \tag{150}$$

If Eqs. (147) and (149) are substituted into inequality (150), one obtains $-c \geq 0$, which contradicts Eq. (148). So the Garg–Mermin model satisfies all inequalities for correlation functions but violates at least this inequality for joint probabilities: then the model cannot be reproduced by a local probabilistic theory.

It can therefore be concluded that the set of inequalities (142) is the widest set of inequalities deduced from local realism for linear combinations of joint probabilities with real coefficients.

3.4.6. The GHZ Formulation of the EPR Paradox

In 1989, Greenberger, Horne, and Zeilinger[26] (GHZ) formulated an interesting new theorem that has attracted wide attention. They considered a system consisting of three mutually well separated and correlated spin-$\tfrac{1}{2}$ particles, in the context of which an incompatibility is demonstrated between quantum mechanics and local realism. Their demonstration, unlike Bell's theorem, concerns only perfect correlations rather than statistical correlations and requires no inequalities.

The state of the three particles is described by the vector

$$|\Psi\rangle \in \mathbb{C}^2 \otimes \mathbb{C}^2 \otimes \mathbb{C}^2$$

which assumes the form

$$|\Psi\rangle = \frac{1}{\sqrt{2}}\{|1, 1, 1\rangle - |-1, -1, -1\rangle\}$$

with respect to the three z-axial bases. Furthermore

$$O_i|\Psi\rangle = +|\Psi\rangle \tag{151}$$

for all three values of i, with the self-adjoint operators

$$O_1 = \sigma_x^1 \otimes \sigma_y^2 \otimes \sigma_y^3$$
$$O_2 = \sigma_y^1 \otimes \sigma_x^2 \otimes \sigma_y^3$$
$$O_3 = \sigma_y^1 \otimes \sigma_y^2 \otimes \sigma_x^3$$

The result of measuring the x component S_x of the spin of any particle can be predicted with certainty by distant measurements of the y components of the other two, and similarly for the result of measuring the y component S_y of any one of them. According to the EPR reality criterion $S_{x,y}^{1,2,3}$ (S_x, S_y components of particles 1,2,3) must be elements of reality having preassigned values ± 1. From the locality condition it follows that these values are independent of the measurements made. So one can write the relations

$$S_x^1 S_y^2 S_y^3 = +1, \qquad S_y^1 S_x^2 S_y^3 = +1, \qquad S_y^1 S_y^2 S_x^3 = +1 \tag{152}$$

bearing (151) in mind. Remembering that $S_{x,y}^{1,2,3} = \pm 1$, one obtains

$$S_x^1 S_x^2 S_x^3 = +1 \tag{153}$$

from (152). It follows, however, from the definitions of O_1, O_2, O_3 that

$$O_1 O_2 O_3 = -\sigma_x^1 \otimes \sigma_x^2 \otimes \sigma_x^3$$

so that from $O_1 O_2 O_3|\Psi\rangle = |\Psi\rangle$ it follows that

$$\sigma_x^1 \otimes \sigma_x^2 \otimes \sigma_x^3|\Psi\rangle = -|\Psi\rangle$$

which contradicts (153). Therefore the predictions derived from $|\Psi\rangle$ are incompatible with local realism.

Mermin[27] considered this "an altogether more powerful refutation of the existence of elements of reality than the one provided by Bell's theorem . . . ." The physical plausibility of state $|\Psi\rangle$ can, however, be questioned; nobody has yet managed to produce it. Furthermore, the argument is strictly restricted to the deterministic form of local realist theories, whereas Bell's theorem holds for probabilistic local realist theories as well.

3.5. EXPERIMENTS WITH ATOMIC PHOTON PAIRS

Several experiments to test the validity of the strong inequalities have been carried out with photons. The quantum-mechanical treatment of photon polarization is similar to that of spin-$\frac{1}{2}$ in the important respect that both observables are dichotomic. In the case of photon polarization this property is actually due to the absence of a photon mass, a fact that has the practical effect of eliminating from the theoretical scheme the photons with longitudinal polarization. Only photons whose linear polarization is perpendicular to the direction of propagation are left, a situation similar to that of classical electromagnetic waves whose transverse nature is well known. In the study of the EPR paradox for atomic photon pairs it is interesting to consider linear and circular polarization.

3.5.1. Correlated Atomic Photon Pairs

In quantum theory one usually defines

$|R\rangle =$ state vector for right-handed circular polarization
$|L\rangle =$ state vector for left-handed circular polarization
$|x\rangle =$ state vector for linear polarization along the x-axis
$|y\rangle =$ state vector for linear polarization along the y-axis

These two sets of vectors are not unrelated. Elementary textbooks on quantum mechanics show that

$$|R\rangle = \frac{1}{\sqrt{2}}\{|x\rangle + i|y\rangle\}$$
$$|L\rangle = \frac{1}{\sqrt{2}}\{|x\rangle - i|y\rangle\}$$

$$(154)$$

The existence of dichotomic observables for photons means that Bell-type inequalities can also be formulated for real photons. Correspondingly, there are also situations where quantum theory gives a description of the polarization of two correlated photons in terms of nonfactorizable state vectors, analogous to the singlet state of two spin-$\frac{1}{2}$ objects, which imply violations of local realism.

In the case of photons the parity quantum number plays an important role and it is necessary to distinguish, for instance, the states $J^P = 0^+$ and $J^P = 0^-$, represented respectively by the state vectors

$$|0+\rangle = \frac{1}{\sqrt{2}} \{|R_\alpha\rangle |R_\beta\rangle + |L_\alpha\rangle |L_\beta\rangle\}$$

$$|0-\rangle = \frac{1}{\sqrt{2}} \{|R_\alpha\rangle |R_\beta\rangle - |L_\alpha\rangle |L_\beta\rangle\}$$

(155)

These states can also be expressed in terms of linear polarization states by using relations (154) for photons α and β. Taking into account the opposite directions of motion of the photons, one obtains

$$|0+\rangle = \frac{1}{\sqrt{2}} \{|x_\alpha\rangle |x_\beta\rangle + |y_\alpha\rangle |y_\beta\rangle\}$$

$$|0-\rangle = \frac{-i}{\sqrt{2}} \{|x_\alpha\rangle |y_\beta\rangle - |y_\alpha\rangle |x_\beta\rangle\}$$

(156)

The basis states with respect to which linear polarization is expressed are arbitrary. Using the rotated x' and y' axes, one obtains results identical to (156) for both states, with x', y' instead of x, y. This property expresses the invariance of zero-angular-momentum states for rotations about the z-axis.

The argument leading to the EPR paradox could easily be repeated for the foregoing two photon states, starting, for instance, from the observation that elements of reality corresponding to polarization measurements along the x- and x'-axes can be assigned to the photon β from measurements performed (or thought to be performed) on photon α. All the inequalities (of weak and strong types) found in previous sections clearly apply also to correlated photon pairs, since they were deduced merely from the assumed dichotomic nature of the measured quantities.

Several experiments performed from 1972 to 1982 checked the validity of these inequalities. They are collected in Table 3.4, where the atom used, the type of cascade, and the wavelengths of the emitted photons are also shown.

TABLE 3.4. Experiments to Test the Strong Inequalities

Reference	Atom	Cascade	λ_1	λ_2
Freedman and Clauser[19]	^{40}Ca	$4p^2\,^1S_0 \to 4p4s\,^1P_1 \to 4s^2\,^1S_0$	5513	4227
Clauser[32]	^{202}Hg	$9\,^1P_1 \to 7\,^3S_1 \to 6\,^3P_0$	5676	4046
Clauser[33]	^{202}Hg	$9\,^1P_1 \to 7\,^3S_1 \to 6\,^3P_0$	5676	4046
Fry and Thompson[34]	^{200}Hg	$7\,^3S_1 \to 6\,^3P_1 \to 6\,^1S_0$	4358	2537
Aspect et al.[35]	^{40}Ca	$4p^2\,^1S_0 \to 4p4s\,^1P_1 \to 4s^2\,^1S_0$	5513	4227
Aspect et al.[16]	^{40}Ca	$4p^2\,^1S_0 \to 4p4s\,^1P_1 \to 4s^2\,^1S_0$	5513	4227
Aspect et al.[28]	^{40}Ca	$4p^2\,^1S_0 \to 4p4s\,^1P_1 \to 4s^2\,^1S_0$	5513	4227

The $(J = 0) \to (J = 1) \to (J = 0)$ cascade of calcium is the most widely used. In this case one can use the strong inequality

$$-1 \le \frac{3R(\varphi)}{R_0} - \frac{R(3\varphi)}{R_0} - \frac{R_1 + R_2}{R_0} \le 0 \qquad (157)$$

which is an immediate consequence of (107), given that all rates R are proportional to the corresponding probabilities ω. In this case the quantum-mechanical predictions based on the state vector $|0+\rangle$ are, for the quantities in (157),

$$\frac{R(\varphi)}{R_0} = \frac{1}{4}\epsilon_+^1 \epsilon_+^2 + \frac{1}{4}\epsilon_-^1 \epsilon_-^2 F_1(\theta) \cos 2\varphi \qquad (158)$$

$$\frac{R_1}{R_0} = \frac{1}{2}\epsilon_+^1, \qquad \frac{R_2}{R_0} = \frac{1}{2}\epsilon_+^2 \qquad (159)$$

where $F_1(\theta)$ is a function of the half-angle θ subtended by the primary lenses representing a depolarization due to noncollinearity of the two photons and where

$$\epsilon_\pm^1 = \epsilon_M^1 \pm \epsilon_m^1, \qquad \epsilon_\pm^2 = \epsilon_M^2 \pm \epsilon_m^2 \qquad (160)$$

Here ϵ_M^1 (ϵ_m^1) is the transmittance of the first polarizer for light polarized parallel (perpendicular) to the polarizer axis; a similar notation has been used for the second polarizer. All these transmittances are usually very near the ideal case, with ϵ_M^i close to 1 and ϵ_m^i close to 0 ($i = 1, 2$). The values of these parameters are collected in Table 3.5 for the indicated experiments. The notation used in Table 3.5 is the usual one, in the sense that index 1 (2) on an optical parameter refers to the first (second) photon of the cascade, with wavelength λ_1 (λ_2) in Table 3.4. One should add two important observations. First, the experiment by Aspect *et al.*[16] used as polarizers two-way polarizing cubes made of two prisms with suitable dielectric thin films on the sides that are stuck together. For such

TABLE 3.5. Optical Transmittances of the Two Polarizers

Reference	ϵ_M^1	ϵ_m^1	ϵ_M^2	ϵ_m^2
Freedman–Clauser 1972[17]	0.97 ± 0.01	0.038 ± 0.004	0.96 ± 0.01	0.037 ± 0.004
Holt–Pipkin 1973[29]	0.910 ± 0.001	$<10^{-4}$	0.880 ± 0.001	$<10^{-4}$
Clauser 1976[32]	≈ 0.965	≈ 0.011	≈ 0.972	≈ 0.008
Fry–Thompson 1976[34]	0.98 ± 0.01	0.02 ± 0.005	0.97 ± 0.01	0.02 ± 0.005
Aspect et al. 1981[35]	0.972 ± 0.005	0.029 ± 0.005	0.968 ± 0.005	0.028 ± 0.005
Aspect et al. 1982[16]	0.950 ± 0.005	0.007 ± 0.005	0.930 ± 0.005	0.007 ± 0.005
Aspect et al. 1982[28]	0.96 ± 0.01	0.005 ± 0.005	0.92 ± 0.01	0.007 ± 0.005

polarizers the notation of Table 3.5 is not the usual one, and one should therefore make the identifications

$$T_1^\| = \epsilon_M^1, \quad T_1^\perp = \epsilon_m^1, \quad T_2^\| = \epsilon_M^2, \quad T_2^\perp = \epsilon_m^2$$

The second observation is that the experiment by Aspect et al.[28] used time-varying analyzers and that each of these switching devices was followed by *two* polarizers in two different orientations. Therefore, the optical parameters for this experiment should actually be doubled for each photon. In practice, however, it happens that the two "large" transmittances are identical for each photon and that the two "small" ones are different by insignificant amounts within the quoted errors. It follows that the parameters in Table 3.5 are good for the two polarizers following every switching device.

As one can see from Table 3.4, some experiments used mercury isotopes and cascades of the type $(J = 1) \rightarrow (J = 1) \rightarrow (J = 0)$. Since in the initial atomic scale there are three levels with different J_3 values, the state of the emitted photon pair is not unique and the calculation is slightly more complicated than in the case of the calcium cascade. The result is simple, however, if it is assumed that the three different initial atomic levels are equally populated and incoherent. One of them obtains for the mercury $(J = 1) \rightarrow (J = 1) \rightarrow (J = 0)$ cascade that

$$\frac{R(\varphi)}{R_0} = \frac{1}{4}\epsilon_+^1\epsilon_+^2 - \frac{1}{4}\epsilon_-^1\epsilon_-^2 F_2(\theta)\cos 2\varphi \tag{161}$$

$$\frac{R_1}{R_0} = \frac{1}{2}\epsilon_+^1, \qquad \frac{R_2}{R_0} = \frac{1}{2}\epsilon_+^2 \tag{162}$$

In (161), $F_2(\theta)$ is a new function of the half-angle θ subtended by the primary lenses and again represents the depolarization due to noncollinearity of the two photons.

3.5.2. Review of Published Experiments

The published experiments on the EPR paradox performed using atomic photon pairs will now be reviewed. It should be borne in mind that no experiment has ever violated a weak inequality.

1. **Freedman and Clauser**. In this experiment the $3d4p\ ^1P_1$ state of calcium in a beam was excited by radiation from a deuterium arc lamp. About 10% of the atoms went to the $4p^2\ ^1S_0$ state, which is the initial state of the $(0, 1, 0)$ cascade emitting photons of wavelengths 551.3 nm and 422.7 nm. See Fig. 3.7. Since the natural calcium used in this experiment was an almost pure sample of the isotope with zero nuclear spin, there was no significant reduction expected in the polarization correlation due to hyperfine structure. On each side of the source

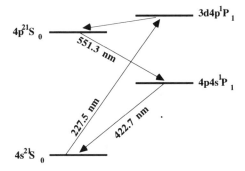

FIGURE 3.7. Level scheme for calcium. The upward-pointing line shows the atomic excitation used for producing the initial level $4p^2\ ^1S_0$, starting point of the atomic cascade in the experiment of Freedman and Clauser.[17]

the photons were collected and collimated by a lens, then passed through a filter and a linear polarizer to a photomultiplier. Freedman and Clauser[17] used pile-of-plates polarizers each of which was about 1 m long and consisted of 10 glass sheets inclined nearly at the Brewster angle.

The photomultiplier pulses were fed to a coincidence circuit, and coincidence measurements were made for 100-s time intervals, the intervals during which all the plates were removed alternating with intervals in which the plates were inserted. The results obtained, as the relative orientation of the transmission axes of the polarizers was varied, were found to be in good agreement with the quantum-mechanical predictions.

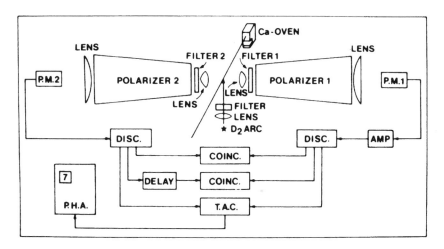

FIGURE 3.8. Schematic of the apparatus of Freedman and Clauser.[17]

The strong inequality (111) can be written

$$\delta \leq 0.250 \qquad (163)$$

if

$$\delta = \left| \frac{R(22\frac{1}{2}^\circ)}{R_0} - \frac{R(67\frac{1}{2}^\circ)}{R_0} \right| \qquad (164)$$

because the ratio of the coincidence rates equals the ratio of the corresponding probabilities. The results for $R(22\frac{1}{2}^\circ)$ and for $R(67\frac{1}{2}^\circ)$, combined with those with both polarizers removed for R_0, gave $\delta = 0.300 \pm 0.008$, in clear violation of Freedman's strong inequality and in agreement with the quanum-mechanical prediction $\delta_{QM} = 0.301 \pm 0.007$.

2. **Holt–Pipkin.**[29] In the second experiment the 567-nm and 404.7-nm photons emitted in the $(1,1,0)$ cascade of the zero-nuclear-spin isotope ^{198}Hg of mercury were observed. Since the final cascade level is not the ground state of the atom, no precautions had to be taken to avoid the effects of resonance trapping observed in the Freedman–Clauser experiment.[30] To produce the required radiation, mercury vapor was excited to the $9\,^1P_1$ state by a 100-eV electron beam, both the beam and the vapor contained in the encapsulated source Pyrex glass. Calcite polarizers were used. These polarizers have a much better extinction ratio than pile-of-plates polarizers, but the values of ϵ_M are somewhat low (Table 3.5).

Experimentally it was found that $\eta = 0.216 \pm 0.013$, a result that disagrees with the quantum-mechanical prediction $\eta_{QM} = 0.266$ and clearly does *not* violate the strong inequality. This discrepancy has never been completely explained. Proponents of local realistic theories have suggested[31] that the use of calcite polarizers may be significant.

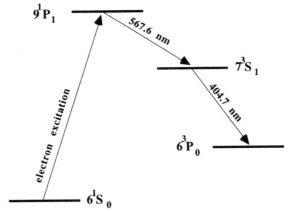

FIGURE 3.9. Level scheme for mercury. The upward-pointing line shows the electron excitation of the initial level $9\,^1P_1$ used in the experiment of Holt and Pipkin[29] as starting point of the atomic cascade.

3. **Clauser.** Clauser[32] repeated the Holt-Pipkin experiment, using the same cascade but in the ^{202}Hg isotope of mercury. Also, instead of calcite polarizers, he used pile-of-plates polarizers. This experiment gave $\delta = 0.2885 \pm 0.0093$, violating Freedman's strong inequality and in close agreement with the quantum-mechanical prediction $\delta_{QM} = 0.2841$.

In an extension of the previous experiment, Clauser[33] measured the *circular* polarization correlation by inserting quarter-wave plates between each linear polarizer and the source. The quarter-wave plates were obtained by applying pressure to bars of commercial-grade quartz. Assuming ideal quarter-wave plates, quantum mechanics predicts that the zero-angular-momentum state vectors (156) should remain unmodified after the two photons have crossed the plates. Therefore, (163) also remains a valid form of Freedman's strong inequality. From the experimental results Clauser found $\delta = 0.235 \pm 0.025$, while, taking into account the transmission efficiencies of the polarizers and the assumed lack of stability of the quarter-wave plates, he obtained from the theory $\delta_{QM} = 0.252$ [which almost does not violate (163)]. Within the limit of experimental errors, these circular-polarization results were in agreement with quantum mechanics, but the whole argument was not very satisfactory.

4. **Fry and Thompson.** These authors[34] used the 435.8-nm and 253.7-nm photons emitted in the $(1,1,0)$ cascade in the zero-nuclear-spin isotope ^{200}Hg of mercury (Table 3.4). The $7\,^3S_1$ state of a mercury beam was populated in a two-step process with electron bombardment excitation of the $6\,^3P_2$ metastable state followed downstream, where all short-lived states had decayed, by absorption of resonant radiation from a tunable dye laser (Fig. 3.10). The laser bandwidth was

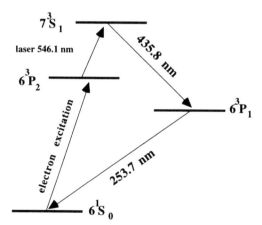

FIGURE 3.10. Level scheme for mercury. The upward-pointing lines show the route for producing the initial level $7\,^3S_1$ used in the experiment of Fry and Thompson[34] as starting point of the atomic cascade.

narrow enough that the ^{200}Hg isotope could be selectively excited. The polarizers used in this experiment were of the pile-of-plates variety, and the earth magnetic field in the interaction volume was reduced to less than 5 mG.

Since the initial state of the cascade had $J = 1$, it was necessary to take into account the possibility of unequal population of, and coherence between, the initial Zeeman sublevels, which Fry and Thomson did by measuring the polarization of the 435.8-nm radiation. Allowing for such effects and considering the transmission efficiencies of Table 3.5, they predicted $\delta_{QM} = 0.294 \pm 0.007$, but experiment gave $\delta = 0.296 \pm 0.014$, in agreement with quantum theory but in violation of Freedman's strong inequality.

5. **Aspect et al.** Aspect, Grangier, and Roger[35] (AGR) used the 551.3-nm and 422.7-nm photons from the (0, 1, 0) cascade of calcium. In their case the calcium atoms were excited to the $4p^2\ ^1S_0$ by a nonresonant two-photon absorption process, using a krypton–ion laser beam of wavelength 406 nm and a dye laser beam tuned to 581 nm, both laser beams being at right angles to the calcium atomic beam emitted from a tantalum oven. The laser beams were focused at the interaction region to provide a source about 60 μ in diameter by 1 mm long. The density varied between $3 \times 10^{10}\,\mathrm{cm}^{-3}$ and $10^{10}\,\mathrm{cm}^{-3}$, resulting in cascade rates equal to or higher than $4 \times 10^7\,\mathrm{s}^{-1}$. Selective excitation of the ^{40}Ca isotope of calcium prevented the polarization correlation from being reduced by hyperfine-

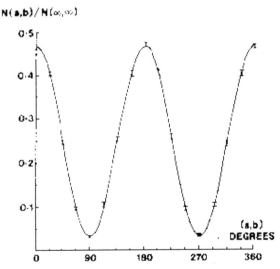

FIGURE 3.11. Normalized coincidence rate as a function of the relative orientation of the transmission axes of the polarizers in the experiment of Aspect et al.[35] The solid curve is the quantum-theoretical prediction.

structure effects. The photons from the cascade were analyzed by polarizers of the pile-of-plates type and filters in much the same way as in previous experiments.

The high atomic density of the source generated coincidence rates up to $100 \, s^{-1}$, allowing a 1% statistical accuracy in only 100 s counting time. Measuring $R(22\frac{1}{2}^\circ)$, $R(67\frac{1}{2}^\circ)$, and R_0, AGR obtained $\delta = 0.3072 \pm 0.0043$, in agreement with the quantum-mechanical prediction of $\delta_{QM} = 0.308 \pm 0.002$ but in violation of Freedman's (strong) inequality by more than 13 standard deviations.

6. **Aspect** *et al.* In 1982, AGR[16] performed an experiment, originally suggested and analyzed by Garuccio and Rapisarda,[15] using two-channel polarizers instead of the previous one-channel pile-of plates types. Each polarizer was a polarizing cube, built using the properties of dielectric thin films and antireflection coated and was rotatable about the observation axis. The arrangement allowed the quantity $E(a, b)$ defined in Eq. (95) to be measured directly in a single run, using a fourfold coincidence technique for each of the four relative orientations of the polarizers: (a, b), (a, b'), (a', b), and (a', b').

If the left side of the (strong) inequality (103) is called S, AGR found experimentally that $S = 2.697 \pm 0.015$, in violation of (103) itself but in full agreement with the quantum-mechanical prediction $S_{QM} = 2.70 \pm 0.05$.

7. **Aspect** *et al.* In all experiments described so far the transmission axes of the polarizers were held fixed during every set of measurements. Thus, in principle,

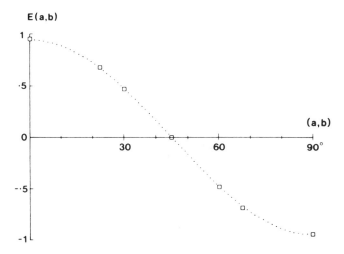

FIGURE 3.12. Correlation function $E(a, b)$ as a function of the relative angle between the transmission axes of the polarizers in the experiment by Aspect *et al.*[16] The squares are the experimental points with error included. The dashed curve is the quantum-theoretical prediction.

there was the possibility of an exchange of information between the two polarizers with a velocity not exceeding that of light. Such a possibility, although very unlikely given the known nature of interactions could at least, in principle, be ruled out if the settings of the polarizers were changed in a time shorter than the time of flight of the photons from the source to the polarizers. In the experiment performed by Aspect, Dalibard, and Roger[28] (ADR), an optical switch rapidly redirected the light incident from the source to one of two polarizing cubes on each side of the source (Fig. 3.13). In contrast to the previous experiment, only the transmitting channels of the polarizing cubes were used. Switching the light was obtained by a Bragg reflection from an ultrasonic standing wave in water. The light was completely transmitted when the amplitude of the standing wave was zero and was almost fully deflected through 10 mrad when the amplitude was a maximum. Switching between the two channels occurred about once every 10 ns, and since time, as well as the 5-ns lifetime of the intermediate level of the cascade, was smaller than L/c (40 ns), where $L = 12$ m was the separation between the two switches and c is the speed of light, a detection event on one side and the corresponding change of orientation on the other side were separated by a spacelike interval.

In the ADR experiment the coincidence rates were only a few per second, with an accidental background of about 1 per second. If U is the intermediate

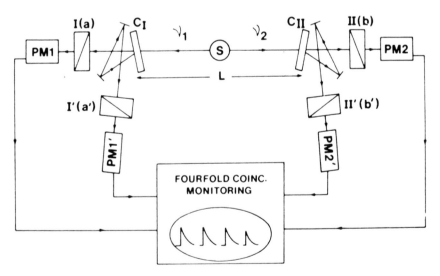

FIGURE 3.13. The ADR experiment with optical switches.[28] The switching devices C_I and C_{II} are followed by differently oriented polarizers. The arrangement was thought to be equivalent to one in which a single polarizer on each side is switched quickly between two different orientations. The distance L was 12 m.

quantity in the (strong) inequality (157), ADR experimentally found $U = 0.101 \pm 0.020$, in clear violation of (157) itself but in agreement with the quantum-mechanical prediction $U_{QM} = 0.112$.

Although in practice the switching was periodic rather than random, the switches on the two sides were driven by two different generators at different frequencies, and it was *assumed* that they functioned in an uncorrelated way. In fact this situation can hide a conceptual difficulty as was shown by Zeilinger.[36]

Finally it should be noted that some criticism[37] of the AGR and ADR experiment has been raised on the grounds that there may have been significant resonance trapping due to the high density of the calcium source. A reply to these criticisms was given by Aspect and Grangier.[38]

8. **Perrie** *et al.* Perrie, Duncan, Beyer, and Kleinpoppen[39] (PDBK) measured for the first time the polarization correlation of the two photons emitted simultaneously by metastable atomic deuterium in a true second-order decay process. Single-photon decay from the $2S_{1/2}$ state of deuterium is forbidden, and the main channel for the deexcitation is by the *simultaneous* emission of two photons that can have any wavelength consistent with energy conservation for the pair. However, because of the absorption in oxyen, the observation window in practice was limited between 185 and 355 nm.

In the PDBK experiment a 1-keV metastable atomic deuterium beam of density about $10^4 \, \text{cm}^{-3}$ was produced by charge exchange, in cesium vapor, of deuterons extracted from a radio-frequency ion source. Electric field prequench plates upstream from the observation region allowed the $2S_{1/2}$ component of the beam to be switched on and off by Stark-mixing the $2S_{1/2}$ and $2P_{1/2}$ states, and at the end of the apparatus the beam was fully quenched so that the resulting Lyman signal could be used to normalize the two-photon coincidence signal. The two-photon radiation was collected and collimated by a pair of lenses, and the polarizers were of the pile-of-plates type.

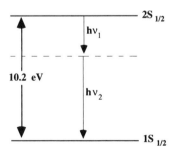

FIGURE 3.14. Level diagram for atomic deuterium, used in the experiment of Perrie *et al.*[39] Hyperfine structure has been neglected. Photonic frequencies v_1 and v_2 can have any value, provided only that $hv_1 + hv_2 = 10.2$ eV, where h is Planck's constant.

Measuring $R(22\frac{1}{2}^{\circ})$, $R(67\frac{1}{2}^{\circ})$, and R_0, PDBK obtained $\eta = 2.268 \pm 0.010$, in agreement with the quantum-theoretical prediction $\eta_{QM} = 0.272 \pm 0.008$, but in violation of Freedman's strong inequality by slightly more than two standard deviations.

In an extension of this last experiment,[40] the circular-polarization correlation was measured by placing achromatic quarter-wave plates in each detection arm between linear polarizer and source. The obtained results did *not* violate Freedman's inequality and would have disagreed with the quantum-theoretical predictions *if* the quarter-wave plates had been assumed to be perfectly achromatic. With the introduction of a considerable degree of achromaticity, PDBK could reconcile their observations with theory, considering also imperfect parallelism of the incoming photons. It is surprising, however, that the only two measurements of circular polarizations in EPR experiments (the present one and the already discussed Clauser experiment[33]) led to results difficult to reconcile with quantum theory.

3.5.3 Empirical Falsification of the Furry Hypothesis

We shall now show that quantum-mechanical factorizable vectors for atomic photon pairs always satisfy Bell-type inequalities, not only of the weak but also of the strong kind. Consider a typical EPR experiment performed with polarizers and detectors, using a source producing correlated (α,β)-pairs of photons. Observables corresponding to the action of the polarizers are represented by $A(\hat{a})$ and $B(\hat{b})$, where $\hat{a}$ and $\hat{b}$ are unit vectors in the x–y plane, that denote the directions of the transmission axes of the polarizers on the left (photon α) and right (photon β). Eigenvalues of A, B are ± 1, where ± 1 corresponds to photons having polarization parallel to the transmission axis, while -1 corresponds to those having polarization perpendicular to the transmission axis. Let the eigenvectors of A and B be $|a\pm\rangle$ and $|b\pm\rangle$, respectively. Then, in terms of the linear polarization basis vectors $|x\rangle$ and $|y\rangle$,

$$
\begin{aligned}
|a+\rangle &= \cos\theta_\alpha|x_\alpha\rangle + \sin\theta_\alpha|y_\alpha\rangle \\
|a-\rangle &= -\sin\theta_\alpha|x_\alpha\rangle + \cos\theta_\alpha|y_\alpha\rangle \\
|b+\rangle &= \cos\theta_\beta|x_\beta\rangle + \sin\theta_\beta|y_\beta\rangle \\
|b-\rangle &= -\sin\theta_\beta|x_\beta\rangle + \cos\theta_\beta|y_\beta\rangle
\end{aligned}
\tag{165}
$$

where $|x_\alpha\rangle$ and $|y_\alpha\rangle$ denote polarization states on the left, $|x_\beta\rangle$ and $|y_\beta\rangle$ on the right, and θ_α and θ_β are the angles between the x-axis and $\hat{a}$ and $\hat{b}$ respectively. The self-adjoint operators $A(\hat{a})$ and $B(\hat{b})$ satisfy

$$
\begin{aligned}
A(\hat{a})|a\pm\rangle &= (\pm 1)|a\pm\rangle \\
B(\hat{b})|b\pm\rangle &= (\pm 1)|b\pm\rangle
\end{aligned}
\tag{166}
$$

Joint probabilities of detections (after passage through the polarizers) can then be written as (perfect polarizers are considered for simplicity):

$$\omega(a+, b+) = |\langle(b+)(a+)|\Psi\rangle|^2 \eta_a \eta_\beta \tag{167}$$

where η_α and η_β are the efficiencies of the photodetectors on the left and right, respectively, and satisfy

$$0 < \eta_\alpha, \eta_\beta \leq 1 \tag{168}$$

Take, for instance,

$$|\Psi\rangle = |u\rangle|v\rangle \tag{169}$$

as a factorizable state vector of the photon pair with $|u\rangle$ and $|v\rangle$ describing photon α and photon β, respectively. Since one can always write

$$\begin{aligned} |u\rangle &= \cos\varphi_\alpha|x_\alpha\rangle + \sin\varphi_\alpha|y_\alpha\rangle \\ |v\rangle &= \cos\varphi_\beta|x_\beta\rangle + \sin\varphi_\beta|y_\beta\rangle \end{aligned} \tag{170}$$

from (165) and (170), it follows that

$$\langle a+|u\rangle = \cos(\theta_\alpha - \varphi_\alpha), \qquad \langle b+|v\rangle = \cos(\theta_\beta - \varphi_\beta) \tag{171}$$

which leads to

$$\omega(a+, b+) = D_\alpha(a+)D_\beta(b+) \tag{172}$$

with

$$\begin{aligned} D_\alpha(a+) &= \cos^2(\theta_\alpha - \varphi_\alpha)\eta_\alpha \\ D_\beta(b+) &= \cos^2(\theta_\beta - \varphi_\beta)\eta_\beta \end{aligned} \tag{173}$$

Obviously

$$0 \leq D_\alpha(a+) \leq \eta_\alpha, \qquad 0 \leq D_\beta(b+) \leq \eta_\beta \tag{174}$$

Next use the Clauser–Horne lemma (72)–(73), from which one can deduce physically meaningful inequalities by making the identifications

$$\begin{aligned} x &= D_\alpha(a+), & x' &= D_\alpha(a'+) \\ y &= D_\beta(b+), & y' &= D_\beta(b'+) \end{aligned} \tag{175}$$

The inequalities (174) being satisfied if a', b' replace a, b, respectively, one gets, by comparing (71)–(72) with (174) and using (175),

$$X = \eta_\alpha, \qquad Y = \eta_\beta \tag{176}$$

Inequalities (73) then become

$$-\eta_\alpha\eta_\beta \leq \Gamma - D_\alpha(a'+)\eta_\beta - \eta_\alpha D_\beta(b+) \leq 0 \tag{177}$$

where

$$\Gamma = \omega(a+, b+) - \omega(a+, b'+) + \omega(a'+, b+) + \omega(a'+, b'+) \qquad (178)$$

Define other double-detection probabilities with one or both polarizers removed as follows:

$$\omega(a+, \infty) = D_\alpha(a'+)\eta_\beta$$
$$\omega(\infty, b+) = \eta_\alpha D_\beta(b+) \qquad (179)$$
$$\omega_0 = \eta_\alpha \eta_\beta$$

where ∞ indicates that the corresponding polarizer has been removed. In place of (177) one can now write

$$-\omega_0 + \omega(a'+, \infty) + \omega(\infty, b+) \le \Gamma \le \omega(a'+, \infty) + \omega(\infty, b+) \qquad (180)$$

The later are exactly the *strong inequalities* used in the experimental study of the EPR paradox. We have thus shown that factorizable state vectors necessarily satisfy the strong inequalities derived from local realism in conjugation with *ad hoc* additional assumptions. That such state vectors also satisfy the weak inequalities is well known.

In actual experiments it has been conclusively observed that "strong" inequalities are violated in nature, which implies that a factorizable state vector can never be taken as a valid description in an EPR-type situation.

The foregoing conclusion can also be verified to be true if a set S of correlated quantum systems (α, β) is described by a mixture of factorizable state vectors $|\psi_1\rangle, |\psi_2\rangle, \ldots, |\psi_k\rangle$ given by

$$|\psi_1\rangle = |u_1\rangle|v_1\rangle \qquad \text{with probability } \rho_1$$
$$|\psi_2\rangle = |u_2\rangle|v_2\rangle \qquad \text{with probability } \rho_2$$
$$\vdots \qquad\qquad\qquad\qquad (181)$$
$$|\psi_k\rangle = |u_k\rangle|v_k\rangle \qquad \text{with probability } \rho_k$$

with

$$\rho_1 + \rho_2 + \cdots + \rho_k = 1 \qquad (182)$$

All Eqs. (167)–(180) can be repeated with the insertion of an index i ($1 \le i \le k$) for the subset S_i of S having state vector $|\psi_i\rangle$ and relative population ρ_i. In place of (180), one now gets

$$-\omega_0 + \omega_i(a'+, \infty) + \omega_i(\infty, b+) \le \Gamma_i \le \omega_i(a'+, \infty) + \omega_i(\infty, b+) \qquad (183)$$

where ω_0 is the same as in (179) and

$$\omega_i(a'+, \infty) = D_{i\alpha}(a'+)\eta_\beta$$
$$\omega_i(\infty, b+) = \eta_\alpha D_{i\beta}(b+)$$

(184)

$$\Gamma_i = \omega_i(a+, b+) - \omega_i(a+, b'+) + \omega_i(a'+, b+) + \omega_i(a'+, b'+)$$

(185)

with

$$\omega_i(a+, b+) = D_{i\alpha}(a+)D_{i\beta}(b+)$$

(186)

and so on. In (184) and (186) one has

$$D_{i\alpha}(a+) = \cos^2(\theta_\alpha - \varphi_{i\alpha})\eta_\alpha$$
$$D_{i\beta}(b+) = \cos^2(\theta_\beta - \varphi_{i\beta})\eta_\beta$$

(187)

where $\theta_\alpha, \theta_\beta$ are given in (165) and $\varphi_{i\alpha}, \varphi_{i\beta}$ are defined by the equations

$$|u_i\rangle = \cos\varphi_{i\alpha}|x_\alpha\rangle + \sin\varphi_{i\alpha}|y_\alpha\rangle$$
$$|v_i\rangle = \cos\varphi_{i\beta}|x_\beta\rangle + \sin\varphi_{i\beta}|y_\beta\rangle$$

(188)

The strong inequality (183) is a consequence of the Clauser–Horne lemma (72)–(73), of identifications similar to (175) with the extra index i for the ω functions, and of the inequalities (174), which are satisfied by $D_{i\alpha}$ and $D_{i\beta}$ as well, as one can immediately check from (187).

All the double-detection probabilities for S can be obtained as weighted averages of the corresponding probabilities for the subsets S_i. By writing

$$\bar{\omega}_\alpha(a'+, \infty) = \sum_{i=1}^{k} \rho_i D_{i\alpha}(a'+, \infty)$$

$$\bar{\omega}_\beta(\infty, b+) = \sum_{i=1}^{k} \rho_i D_{i\beta}(\infty, b+)$$

(189)

$$\bar{\omega}(a+, b+) = \sum_{i=1}^{k} \rho_i D_i(a+, b+)$$

one gets, by direct averaging of (183) and (185),

$$-\omega_0 + \bar{\omega}(a'+, \infty) + \bar{\omega}(\infty, b+) \le \bar{\Gamma} \le \bar{\omega}(a'+, \infty) + \bar{\omega}(\infty, b+)$$

(190)

where

$$\bar{\Gamma} = \bar{\omega}(a+, b+) - \bar{\omega}(a+, b'+) + \bar{\omega}(a'+, b+) + \bar{\omega}(a'+, b'+)$$

(191)

Obviously (190)–(191) is the strong inequality for the mixture (181). Its validity for all conceivable mixtures of factorizable state vectors means that the hope entertained in the past by several important authors (e.g., see Jauch,[41] de Broglie,[42] and Piccioni[43]) that the solution of the EPR paradox could be found in spontaneous factorization of quanum-mechanical "entangled" state vectors is actually untenable, because experiments have shown conclusively that the strong inequality is violated.

3.5.4. Inequality for Parametric Down-Conversion Experiments

In this section it will be shown that the factorizable state vector produced by a beam splitter placed on the paths of parametric down-converted photons violates some inequalities, deduced from local realism, that are more stringent than those normally used in the discussion of quantum nonlocality.

In the last few years the parametric down-conversion of light has been widely used in quantum optics. The signal and idler photons produced in a nonlinear crystal are highly correlated in momentum and in polarization. The setups used in the EPR parametric down-conversion experiments performed up to now are very different[44–50] but most share the feature that the state vector used to test quantum mechanics is factorizable. The most important consequence of the quantum-mechanical description of correlations with nonfactorizable ("entangled") state vectors is of course the possibility of violating local realism. The use of factorizable state vectors, instead of entangled ones, can therefore be questioned as a proper way to test local realism versus quantum mechanics.[51,52] The reason, seen in Section 3.2.2, is that every factorizable state vector satisfies Bell's inequality. This is technically correct, of course, but it can be shown[53] that an inequality very similar to Bell's can be proved with an upper limit 1 (instead of 2), which is violated by the consequences of the factorizable state vector relevant to the parametric down-conversion experiments.

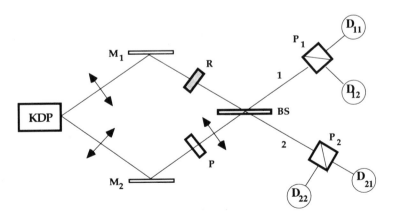

FIGURE 3.15. Setup of an EPR experiment with pairs of photons from parametric down-conversion in a KDP crystal. Double arrows indicate states of polarization. Photons are reflected on the mirrors M_1 and M_2 and cross the beam splitter BS. The two-way polarizer P_1 (P_2) splits the photon trajectory into two, on which are placed the photon detectors D_{11} and D_{12} (D_{21} and D_{22}).

The section has three aims: (i) to apply local realism to photon pairs produced in parametric down-conversion and obtain, for this case, a new inequality more stringent than the conventional Bell inequality; (ii) to prove that the relevant quantum-mechanical state vector, even if factorizable, violates this new inequality; (iii) to stress that also in this case the supplementary assumptions play an essential role in the published experimental attempts to discriminate between local realism and quantum mechanics. The arguments of Ref. 53 will be followed, with the conclusion that parametric down-conversion experiments can be used to compare the predictions of local realism with the existing quantum theory if and only if the new inequality given here, and not the conventional one, is applied.

Since the conceptual situation is very similar in most of the mentioned setups, attention will be focused on the experimental setting used in Ref. 16 and shown in Fig. 3.15. Extension to the other situations is straightforward. A laser beam impinges on a nonlinear type I crystal, e.g., of potassium dihydrogen phosphate. Laser photons are down-converted to pairs of photons with the same polarization, which propagate in different directions. The special pairs of the degenerate case (same frequency) are selected, and the polarization of one of the photons is rotated 90° by a wave plate. The two photons, after reflecting on two mirrors, are directed to the opposite sides of the polarization-independent beam splitter (Fig. 316). The state vector of the emerging pair is

$$|\psi\rangle = \left(\sqrt{T_x}|x_1\rangle + i\sqrt{R_x}|x_2\rangle\right) \otimes \left(\sqrt{T_y}|y_2\rangle - i\sqrt{R_y}|y_1\rangle\right) \tag{192}$$

which is clearly a tensor product. Here $|x_i\rangle$ $[|y_i\rangle]$ is the polarization state along the x $[y]$ direction for the photon in the ith output channel of the beam splitter ($i = 1, 2$), and $R_x, R_y, T_x,$ and T_y are the reflectivities and transmittivities of the beam splitter.

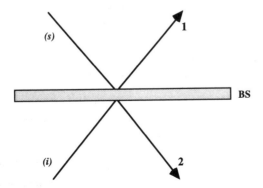

FIGURE 3.16. Signal (s) and idlder (i) photons arising in parametric down-conversion processes are directed toward the opposite sides of a beam splitter BS. The emerging beams **1** and **2** are described (in part) by a quantum-mechanical two-photon entangled state.

After reorganising terms the state vector can be written in the form

$$|\psi\rangle = \sqrt{T_x T_y}|x_1 y_2\rangle + \sqrt{R_x R_y}|x_2 y_1\rangle - i\sqrt{T_x R_y}|x_1 y_1\rangle + i\sqrt{R_x T_y}|x_2 y_2\rangle \qquad (193)$$

which represents the four possibilities for the two photons that impinge on the beam splitter: signal and idler transmitted, signal and idler reflected, signal transmitted and idler reflected, and signal reflected and idler transmitted. In the case of a 50–50 beam splitter the square-root coefficients appearing in (193) are all 1/2, and the state vector therefore becomes

$$|\psi\rangle = \tfrac{1}{2}[|x_1 y_2\rangle + |x_2 y_1\rangle - i|x_1 y_1\rangle + i|x_2 y_2\rangle)] \qquad (194)$$

We have the following predictions if perfect detectors are used:

[P1] In 25% of the cases there are no counts on path 1 because the component $|x_2 y_2\rangle$ of $|\psi\rangle$ represents two photons present on path 2.

[P2] In another 25% of the cases there are no counts on path 2 because the component $|x_1 y_1\rangle$ of $|\psi\rangle$ represents two photons present on path 1.

[P3] In the remaining 50% of the cases there are always coincident counts on paths 1 and 2 owing to the presence in $|\psi\rangle$ of the $|x_1 y_2\rangle$ and $|x_2 y_1\rangle$ components.

[P4] As a consequence of [P1]–[P3] the probability of a double nondetection on the two paths is zero.

These predictions are clearly nonparadoxical and therefore fully acceptable in the local realistic approach. In fact, all the results obtained here are based on [P1]–[P4]. In particular, a more stringent upper limit for Bell's inequality will be obtained, a result possible only because the specific situation represented by [P1]–[P4] is faced. The new bound does not hold in the general case, for which Bell's inequality is perfectly adequate.

The problem of attributing (local) elements of reality to a statistical ensemble, given its quantum state, can be solved by using the commonly accepted EPR "reality criterion": given a quantum state for a system, if it is possible to predict with certainty (and without perturbing the system) the value of a physical quantity, then there exists an element of reality λ belonging to the system, which corresponds to that physical quantity.

The physical quantity to be considered in the present case is *photon observability* on paths 1 and/or 2 of Fig. 3.16. To start with, one should realize that the state vector (194) is totally indistinguishable from the mixture

$$|\psi_0\rangle = [|x_1 y_2\rangle + |x_2 y_1\rangle)]/\sqrt{2} \quad \text{in 50\% of the cases}$$
$$|\psi_1\rangle = |x_1 y_1\rangle \qquad\qquad\qquad \text{in 25\% of the cases} \qquad (195)$$
$$|\psi_2\rangle = |x_2 y_2\rangle \qquad\qquad\qquad \text{in 25\% of the cases}$$

because the vectors $|\psi_0\rangle$, $|\psi_1\rangle$, and $|\psi_2\rangle$, considered as parts of $|\psi\rangle$, can never interfere with one another when the paths are spatially separated after the interaction with the beam splitter. Therefore, all conclusions concerning the physical situation described by (195) apply to (194) as well.

One can predict with certainty that *perfect* detectors placed on paths 1 and 2 (with or without polarizers) will obtain coincident detections in that 50% of the cases described by $|\psi_0\rangle$. Applying the reality criterion, one concludes that elements of reality λ_0 belong to 50% of the photon pairs, as a consequence of which these pairs have photon observability on both paths.

One can also predict with certainty that perfect detectors placed on path 2 (with or without polarizer) will detect nothing in that 25% of the cases described by $|\psi_1\rangle$. Applying the reality criterion, one concludes that new elements of reality λ_1 belong to 25% of the photon pairs, as a consequence of which these pairs have no observability on path 2.

Finally, one can predict with certainty that perfect detectors placed on path 1 (with or without polarizer) will detect nothing in that 25% of the cases described by $|\psi_2\rangle$. Applying once more the reality criterion, one concludes that new elements of reality λ_2 belong to 25% of the photon pairs, and hence these pairs have no observability on path 1.

Given that $|\psi_0\rangle$, $|\psi_1\rangle$, and $|\psi_2\rangle$ have different physical properties, in the local realistic approach they are described by different elements of reality, that is, by values of $\lambda \in \Lambda$ belonging to different subsets of Λ:

$$|\psi_0\rangle \Rightarrow \lambda = \lambda_0 \in \Lambda_0$$
$$|\psi_1\rangle \Rightarrow \lambda = \lambda_1 \in \Lambda_1 \quad\quad (196)$$
$$|\psi_2\rangle \Rightarrow \lambda = \lambda_2 \in \Lambda_2$$

Naturally, the different values of λ are expected to appear at random, as occurs in general in quantum mixtures. The overlap of any two of these subsets is zero, and their union must be the entire set: $\Lambda = \Lambda_0 \cup \Lambda_1 \cup \Lambda_2$. If $\rho(\lambda)$ is the probability density, the normalization condition must hold:

$$\int_\Lambda d\lambda \, \rho(\lambda) = 1 \quad\quad (197)$$

From (195) it follows that *a priori* probabilities of finding a given value of λ in Λ_0, Λ_1, or Λ_2 are respectively given by

$$\int_{\Lambda_0} d\lambda \, \rho(\lambda) = \frac{1}{2}; \quad \int_{\Lambda_1} d\lambda \, \rho(\lambda) = \frac{1}{4}; \quad \int_{\Lambda_2} d\lambda \, \rho(\lambda) = \frac{1}{4} \quad\quad (198)$$

which are consistent with (197). Naturally Λ_0, Λ_1 and Λ_2 are totally independent of the usual detection parameters (introduced later). For *perfect detectors*, the choice of a photon to remain undetected cannot be made when it interacts locally with a polarizer, because a nonzero probability of joint nondetection would exist, in contradiction to the nonparadoxical quantum-mechanical prediction [P4].

Deterministic Proof

The hidden variable λ is assumed to fix also the nonzero values of Bell's observables: Therefore we write $A(a, \lambda)$ and $B(b, \lambda)$ for the trichotomic ($= \pm 1, 0$) observables measured on the two photons, assuming that 0 means no detection (in coincidence with a detection on the other channel).

In principle, "anomalous" values of $A(a, \lambda)$ and $B(b, \lambda)$ should be considered when $\lambda \in \Lambda_1, \Lambda_2$, since (194) predicts double counts to appear on path 1 [path 2] when $\lambda \in \Lambda_1$ [$\lambda \in \Lambda_2$]. But these one-sided double counts, however represented in the formalism, would in all cases be multiplied by the value 0 of $B(b, \lambda)$ [$A(a, \lambda)$]. In fact, when $\lambda \in \Lambda_1, \Lambda_2$, no photon is detected on one of the two paths, a situation formally represented by

$$A(a, \lambda) = A(a', \lambda) = 0 \qquad \text{if } \lambda \in \Lambda_2$$

and

$$B(b, \lambda) = B(b', \lambda) = 0 \qquad \text{if } \lambda \in \Lambda_1$$

Consequently, we will ignore such anomalous values since they do not contribute to the correlation function of A and B, which is

$$P(a, b) = \int_\Lambda d\lambda \; \rho(\lambda) A(a, \lambda) B(b, \lambda) \qquad (199)$$

Therefore, definition (199) can be rewritten with Λ replaced by Λ_0. After doing this the observables $A(a, \lambda)$ and $B(b, \lambda)$ are such that when a detector gives a result, this can only be ± 1. Therefore,

$$|A(a, \lambda)| = |A(a', \lambda)| = |B(b, \lambda)| = |B(b', \lambda)| = 1 \qquad (200)$$

It is a simple matter to show that

$$|P(a, b) - P(a, b') + P(a', b) + P(a', b')|$$
$$\leq \int_{\Lambda_0} d\lambda \; \rho(\lambda) [|A(a, \lambda)| \, |B(b. \lambda) - B(b', \lambda)|$$
$$+ |A(a', \lambda)| \, |B(b, \lambda) + B(b', \lambda)|] \qquad (201)$$

so that from Eq. (200) one gets

$$|P(a, b) - P(a, b') + P(a', b) + P(a', b')| \leq 2 \int_{\Lambda_0} d\lambda \; \rho(\lambda) \qquad (202)$$

because

$$|B(b, \lambda)| = |B(b', \lambda)| = 1 \Rightarrow |B(b, \lambda) - B(b', \lambda)| + |B(b, \lambda) + B(b', \lambda)| = 2$$

By virtue of (198), inequality (202) becomes

$$|P(a, b) - P(a, b') + P(a', b) + P(a', b')| \le 1 \tag{203}$$

which is Bell's inequality for the physical situation described by the vector (193). The restrictions coming from local realism are now expressed by an upper limit in (203) that is 50% lower than the well-known limit 2 of the traditional form of Bell's theorem.

Probabilistic Proof

Probabilistic local realism was formulated in previous sections, and it was shown that the joint probabilities of the two quantum systems satisfy a factorizability condition with a the probability density ρ depending, in general, on a discrete index λ and on the chosen observables: $\rho = \rho_\lambda(a, a', b, b')$. In place of Eq. (197) one has now

$$\sum_{\lambda \in \Lambda_0} \rho_\lambda(x) = \frac{1}{2}; \quad \sum_{\lambda \in \Lambda_1} \rho_\lambda(x) = \frac{1}{4}; \quad \sum_{\lambda \in \Lambda_2} \rho_\lambda(x) = \frac{1}{4} \tag{204}$$

where $x = \{a, a', b, b'\}$.

Let $p_+(a, \lambda)$ $[p_-(a, \lambda)]$ be the detection probability of the first photon on the ordinary [extraordinary] path outgoing from the first polarizer with axis a and belonging to the λth homogeneous subset, and let $q_\pm(b, \lambda)$ be similar probabilities of the second photon outgoing from a polarizer with axis b. If $p_+(\infty, \lambda)$ and $q_+(\infty, \lambda)$ are the detection probabilities of the first and second photon, respectively, when the polarizers are removed, for ideal instruments one has

$$p_+(a, \lambda) + p_-(a, \lambda) = p_+(\infty, \lambda) \tag{205}$$

and

$$q_+(b, \lambda) + q_-(b, \lambda) = q_+(\infty, \lambda) \tag{206}$$

From [P1]–[P3] it follows that $p_+(\infty, \lambda) = 1, 1, 0$ if $\lambda \in \Lambda_0, \Lambda_1, \Lambda_2$, respectively, and that $q_+(\infty, \lambda) = 1, 0, 1$ if $\lambda \in \Lambda_0, \Lambda_1, \Lambda_2$, respectively. Thus

$$p_+(\infty, \lambda)q_+(\infty, \lambda) = 0 \qquad \text{if } \lambda \in \Lambda_1 \cup \Lambda_2 \tag{207}$$

For the overall joint detections one can write:

$$\omega(a\pm, b\pm) = \int_\Lambda d\lambda \, \rho_\lambda(x)p_\pm(a, \lambda)q_\pm(b, \lambda) \tag{208}$$

Defining the correlation function, as usual, as

$$P(a, b) = \omega(a+, b+) - \omega(a+, b-) - \omega(a-, b+) + \omega(a-, b-) \qquad (209)$$

one gets

$$P(a, b) = \int_\Lambda d\lambda \, \rho_\lambda(x) p(a, \lambda) q(b, \lambda) \qquad (210)$$

where

$$p(a, \lambda) = p_+(a, \lambda) - p_-(a, \lambda) \quad \text{and} \quad q(b, \lambda) = q_+(b, \lambda) - q_-(b, \lambda) \qquad (211)$$

Notice that from (205) and (206) it follows that

$$|p(a, \lambda)| \leq p_+(\infty, \lambda) \quad \text{and} \quad |q(b, \lambda)| \leq q_+(\infty, \lambda) \qquad (212)$$

Naturally (212) holds for all possible a and b. Therefore,

$$|P(a, b) - P(a, b') + P(a', b) + P(a', b')|$$
$$\leq \sum_{\lambda \in \Lambda} \rho_\lambda(x)[|p(a, \lambda)| \, |q(b, \lambda) - q(b', \lambda)| + |p(a', \lambda)| \, |q(b, \lambda) + q(b', \lambda)|]$$
$$\leq \sum_{\lambda \in \Lambda} \rho_\lambda(x) p_+(\infty, \lambda)[|q(b, \lambda) - q(b', \lambda)| + |q(b, \lambda) + q(b', \lambda)|] \qquad (213)$$

the last step being a consequence of (212). Two real numbers $q(b, \lambda)$ and $q(b', \lambda)$ satisfying (212) give

$$|q(b, \lambda) - q(b', \lambda)| + |q(b, \lambda) + q(b', \lambda)| \leq 2q_+(\infty, \lambda) \qquad (214)$$

as one can easily show. Therefore,

$$|P(a, b) - P(a, b') + P(a', b) + P(a', b')| \leq 2\omega(\infty, \infty) \qquad (215)$$

where

$$\omega(\infty, \infty) = \sum_{\lambda \in \Lambda_0} \rho_\lambda(x) p_+(\infty, \lambda) q_+(\infty, \lambda) \qquad (216)$$

is the joint detection probability when both polarizers have been removed. In (216) the sum is taken over $\lambda \in \Lambda_0$ only, because of (207). But in Λ_0 one has $p_+(\infty, \lambda) = q_+(\infty, \lambda) = 1$, so

$$\omega(\infty, \infty) = \sum_{\lambda \in \Lambda_0} \rho_\lambda(x) = \tfrac{1}{2} \qquad (217)$$

because of (204), and inequality (215) is then identical to (203).

Inequalities in the CHSH Form

It is possible to set the new inequality (203) in a form similar to that given in 1969 by CHSH. In fact, using (205) and (206) in (208) one has

$$\omega(a+, b+) + \omega(a+, b-) = \omega(a+, \infty), \quad \omega(a+, b+) + \omega(a-, b+) = \omega(\infty, b+)$$
$$(218)$$

where

$$\omega(a+, \infty) = \sum_{\lambda \in \Lambda_0} \rho_\lambda(x) p_+(a, \lambda) q_+(\infty, \lambda)$$
$$\omega(\infty, b+) = \sum_{\lambda \in \Lambda_0} \rho_\lambda(x) p_+(\infty, \lambda) q_+(b, \lambda) \tag{219}$$

Furthermore, one can easily obtain

$$\omega(a-, b-) = \omega(a+, b+) - \omega(a+, \infty) - \omega(\infty, b+) + \omega(\infty, \infty) \tag{220}$$

By inserting (218) and (220) in (209) one gets

$$P(a, b) = 4\omega(a+, b+) - 2\omega(a+, \infty) - 2\omega(\infty, b+) + \omega(\infty, \infty) \tag{221}$$

which contains only joint detection probabilities on the ordinary rays, also useful in the case of one-way polarizers. The latter equation, inserted in (215), leads to

$$\frac{-1 - 2\omega(\infty, \infty)}{4} \leq B \leq \frac{1 - 2\omega(\infty, \infty)}{4} \tag{222}$$

where

$$B = \omega(a+, b+) - \omega(a+, b'+) + \omega(a'+, b+)$$
$$+ \omega(a'+, b'+) - \omega(a'+, \infty) - \omega(\infty, b+) \tag{223}$$

Remembering (217), Eq. (222) becomes

$$-\tfrac{1}{2} \leq B \leq 0 \tag{224}$$

This inequality can be shown to be violated by the quantum-mechanical predictions following from the state vector (194).

It is important to remember that the previous results are strictly related to the initial assumptions of perfect detection in the apparatus. Indeed if one supposes, as in the actual experiments, that the detection process has a quantum efficiency $\eta \leq 1$, one can show that the upper limit of (224) becomes larger than zero. In fact, the quantum-mechanical joint detection probabilities are in this case all multiplied by η^2; i.e., quantum mechanics assumes that the subset of photons generating coincidences arising from a random process independent of the correlation state of the pair. Therefore, if the same assumption is made in the local realistic approach, one must write

$$\omega(\infty, \infty) = \tfrac{1}{2}\eta^2 \tag{225}$$

so that (222) becomes

$$\frac{-1 - \eta^2}{4} \le B \le \frac{1 - \eta^2}{4} \tag{226}$$

where the lower and upper limits depend on the quantum efficiency η.

In conclusion, it has been shown that in some situations it is possible to obtain a more stringent Bell-type inequality, which is violated by the factorizable state produced in a type I parametric down-conversion source. Even in this case, the low quantum efficiency of the detectors used in the performed experiments does not allow one to draw any conclusion concerning the violation of local realism.

REFERENCES

1. D. BOHM and B. J. HILEY, *Found. Phys.* **5**, 93–109 (1975).
2. B. J. HILEY, *Contemp. Phys.* **18**, 411–414 (1977).
3. P. A. M. DIRAC, in: *Directions in Physics* (H. HORA *et al.*, eds.), Sydney (1976), p. 10.
4. A. SHIMONY, in: *Fundamental Problems in Quantum Theory. A Conference Held in Honor of Professor John. A. Wheeler* (D. M. GREENBERGER *et al.*, eds.), Annals of the New York Academy of Sciences, New York (1995) pp. 675–679.
5. V. CAPASSO, D. FORTUNATO, and F. SELLERI, *Int. J. Theor. Phys.* **7**, 319–326 (1973).
6. L. J. LANDAU, *Phys. Lett. A.* **120**, 54–56 (1987).
7. J. BARRETTO BASTOS FILHO and F. SELLERI, *Found. Phys.* **25**, 701–716 (1995).
8. K. R. POPPER, *A World of Propensities*, Thoemmes, Bristol (1990).
9. J. VON NEUMANN, *Mathematische Grundlagen der Quantenmechanik*, Springer, Berlin (1932).
10. J. F. CLAUSER and M. A. HORNE, *Phys. Rev* **D10**, 526–535 (1974).
11. E. P. WIGNER, *Am. J. Phys.* **38**, 1005–1009 (1970).
12. F. SELLERI, in: *Problems in Quantum Physics: Gdansk '87* (L. KOSTRO, *et al.*, eds.), World Scientific, Singapore (1988), pp. 256–268. See also V. L. LEPORE and F. SELLERI, *Found. Phys. Lett.* **3**, 203–220 (1990).
13. J. F. CLAUSER, M. A. HORNE, A. SHIMONY, and R. A. HOLT, *Phys. Rev. Lett.* **23**, 880–884 (1969).
14. A. ASPECT, *Thèse*, Universitè de Paris-Sud, Centre d'Orsay (1983).
15. A. GARUCCIO and V. RAPISARDA, *Nuovo Cimento* **65A**, 269–297 (1981).
16. A. ASPECT, P. GRANGIER, and G. ROGER, *Phys. Rev. Lett.* **49**, 91–94 (1982).
17. S. J. FREEDMAN and J. F. CLAUSER, *Phys. Rev. Lett.* **28**, 938–941 (1972).
18. F. SELLERI and A. ZEILINGER, *Found Phys.* **18**, 1141–1158 (1988); M. FERRERO and E. SANTOS, *Phys. Lett. A.* **116**, 356–360 (1986); S. PASCAZIO, in: *Quantum Mechanics versus Local Realism, The Einstein–Podolsky–Rosen Paradox* (F. Selleri, ed.) Plenum, New York (1988), pp. 391–411.
19. F. SELLERI, *Found. Phys.* **8**, 103–116 (1978).
20. S. M. ROY and V. SINGH, *J. Phys. A* **11**, 167–171 (1978).
21. A. GARUCCIO and F. SELLERI, *Found. Phys.* **10**, 209–216 (1980).
22. A. GARG, *Phys. Rev. D.* **28**, 785–790 (1983).

23. A. GARUCCIO, in:, *Quantum Mechanics versus Local Realism. The Einstein, Podolsky and Rosen Paradox*, (F. Selleri, ed.) Plenum, New York (1988) pp. 87–113.
24. V. L. LEPORE, *Found. Phys. Lett.* **2**, 15–26 (1989).
25. A. GARG and N. D. MERMIN, *Phys. Rev. Lett.* **49**, 1220–1223 (1982).
26. D. M. GREENBERGER, M. HORNE, and Z. ZEILINGER, in: *Bell's Theorem Quantum Theory and Conceptions of the Universe* (M. KAFATOS, ed.), Kluwer, Dordrecht (1989), pp. 73–76.
27. N. D. MERMIN, *Phys. Today*, pp. 9–11 (June 1990).
28. A. ASPECT, J. DALIBARD, and G. ROGER, *Phys. Rev. Lett.* **49**, 1804–1807 (1982).
29. R. A. HOLT and F. M. PIPKIN, University of Harvard, unpublished preprint (1974).
30. S. J. FREEDMAN, Ph.D. thesis, University of California, Lawrence Berkeley Laboratory Report No. LBL-391 (1972).
31. E. S. CORCHERO, *Proposed Atomic Cascade Experimental Test of Symmetric Local Hidden-Variables Theories*, University of Santander preprint (1986).
32. J. F. CLAUSER, *Phys. Rev. Lett.* **37**, 1223–1226 (1976).
33. J. F. CLAUSER, *Nuovo Cimento* **33B**, 740–746 (1976).
34. E. S. FRY and R. C. THOMPSON, *Phys. Rev. Lett.* **37**, 465–468 (1976).
35. A. ASPECT, P. GRANGIER, and G. ROGER, *Phys. Rev. Lett.* **47**, 460–463 (1981).
36. A. ZEILINGER, *Phys. Lett.* **118A**, 1–2 (1986).
37. F. SELLERI, *Lett. Nuovo Cimento* **39**, 252–256 (1984).
38. A. ASPECT and P. GRANGIER, *Lett. Nuovo Cimento* **43**, 345–348 (1985).
39. W. PERRIE, A. J. DUNCAN, H. J. BEYER, and H. KLEINPOPPEN, *Phys. Rev. Lett.* **54**, 1790–1793 (1985).
40. A. DUNCAN, in: *Book of Invited Papers, Tenth International Conference on Atomic Physics* (H. NAURUMI *et al.*, eds.), Tokyo (1986).
41. L. DE BROGLIE, *C.R. Acad. Sci. Paris* **278**, 721–722 (1974).
42. J. M. JAUCH, in: *Foundations of Quantum Mechanics* (B. d'Espagnat, ed.), Academic Press, New York (1971), pp. 20–55.
43. O. PICCIONI and W. MEHLHOP, in *Microphysical Reality and Quantum Formalism* vol. 1 (A. VAN DER MERWE, *et al.*, eds.), Kluwer, Dordrecht (1988), pp. 375–389.
44. Z. Y. OU and L. MANDEL, *Phys. Rev. Lett.* **61**, 50–53 (1988).
45. C. O. ALLEY and Y. H. SHIH, *Phys. Rev. Lett* **61**, 2921–2924 (1988).
46. T. E. KIESS, Y. H. SHIH, A. V. SERGIENKO, and C. O. ALLEY, *Phys. Rev. Lett.* **71**, 3893–3897 (1993).
47. Z. Y. OU, C. K. HONG, and L. MANDEL, *Opt. Commun.* **67**, 159–163 (1988).
48. S. M. TAN and D. F. WALLS, *Opt. Commun.* **71**, 235–238 (1989).
49. P. G. KWIAT, A. M. STEINBERG, and R. Y. CHIAO, *Phys. Rev.* **A47**, 2472–2475 (1993).
50. J. D. FRANSON, *Phys. Rev. Lett* **62**, 2205–2208 (1989).
51. L. DE CARO and A. GARUCCIO, *Phys. Rev.* **A50**, 2803–2805 (1994).
52. V. CAPASSO, D. FORTUNATO, and F. SELLERI, *Int. J. Theor. Phys.* **7**, 319–326 (1973).
53. A. GARUCCIO, R. RISCO DELGADO, and F. SELLERI, *Phys. Rev. A.* (1998), submitted.

Chapter 4

The EPR Paradox in Particle Physics

A more critical scrutiny of the incompatibility between quantum theory and local realism can come from the study of the EPR paradox in domains where highly efficient particle detectors are used and the additional assumptions are therefore not needed. An appealing possibility is the decay of a $J^{PC} = 1^{--}$ vector meson into a pair of neutral bosons. The copious production of ϕ meson decays into two neutral kaons in a ϕ factory accelerator seems to provide a very useful way of studying the EPR problem (Fig. 4.1). An experiment of this type is characterized by (a) almost perfect angular correlation between the two kaons, (b) nearly 100% efficient high-energy particle detectors, and (c) absence of noise. B factory accelerators also seem to open very interesting new possibilities. These ideas, other proposals, and actually performed experiments on the EPR paradox in nuclear physics are reviewed in this chapter. The discussion will be limited to CP conserving processes, which have a larger probability and seem to allow for easier ways of testing local realism versus quantum mechanics.

4.1. SOME FEATURES OF PARTICLE PHYSICS

In studying the transformations of the subnuclear world, one must bear in mind that all conceivable processes not violating any conservation law do take place, in competition with one another. Among these laws those concerning energy, momentum, angular momentum, and electric charge are well known. The development of subnuclear physics has shown, however, that other conserved "quantum numbers" exist, such as the fundamental baryon number, isotopic spin, strangeness (only approximately conserved but very important), and so on. These quantum numbers, unknown in the macroscopic domain, are the most elementary and fundamental properties of material reality known today. Even if their existence has been irreversibly established, their nature and necessity remains little understood: a task of future research could be to reach a better understanding of them by following the line of thought of local realism.

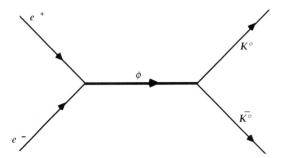

FIGURE 4.1. Feynman diagram for the production of two neutral kaons in electron–positron collisions via an intermediate ϕ meson of mass 1020 MeV/c^2.

4.1.1. Conservation of Energy and Momentum

In high-energy particle collisions it often happens that kinetic energy is transformed into the energy equivalent of a mass. Consider, for example, the process

$$P + P \rightarrow P + P + \pi^0 \tag{1}$$

in which two colliding protons give rise in the final state to two other protons, identical to those of the initial state, and to a π^0 meson, whose mass is created from the "condensation" of part of the kinetic energy of the incoming proton. If $p_i (i = 1, \ldots, 4)$ are the energy–momentum four-vectors of the incoming $(i = 1, 2)$ and outgoing $(i = 3, 4)$ protons, the mass shell conditions

$$p_i^2 = p_i \cdot p_i = \mathbf{p}_i \cdot \mathbf{p}_i - \left(\frac{E_i}{c}\right)^2 = -m^2 c^2 \tag{2}$$

are satisfied, where $\mathbf{p}_i$ are E_i momentum and energy of the ith proton and $m = 938$ MeV/c^2 is the proton rest mass. Equation (2) is a relativistically invariant condition; i.e. $\mathbf{p}_i$ and E_i can be measured in any inertial frame and the result is always the same.

The squared sum of two energy–momentum four-vectors is also invariant. If one denotes quantities calculated in the center-of-mass system with an asterisk, one has

$$(p_1 + p_2)^2 = (\mathbf{p}_1^* + \mathbf{p}_2^*)^2 - \frac{1}{c^2}(E_1^* + E_2^*)^2 = -\frac{1}{c^2}W^2 \tag{3}$$

because $\mathbf{p}_1^* + \mathbf{p}_2^* = 0$, by definition of center of mass, if one writes $W = E_1^* + E_2^*$ for the total energy.

One could—wrongly—believe that the energetic threshold of reaction (1) is given by a kinetic energy of the incoming proton, which is equal to μc^2, where

$\mu = 135$ MeV/c^2 is the neutral pion mass. Actually such a configuration would guarantee the conservation of energy but not that of momentum because the pion and the final protons would be at rest in the laboratory and thus have a total momentum zero, while the initial momentum is considerable when the kinetic energy is μc^2.

The threshold condition must be taken in the center-of-mass system, where the initial total momentum vanishes and is therefore compatible with three final particles at rest in that system. Therefore reaction (1) is possible only if

$$W \geq 2mc^2 + \mu c^2 \tag{4}$$

From (3) it follows that

$$(p_1 + p_2)^2 \leq -\frac{1}{c^2}(2mc^2 + \mu c^2)^2 \tag{5}$$

Now it is possible to use the invariance of the relativistic scalar product and calculate the left side of (5) in the laboratory, where $\mathbf{p}_2 = 0$ and $E_2 = mc^2$. One easily obtains

$$(p_1 + p_2)^2 = -2mT_1 - 4m^2c^2 \tag{6}$$

where $T_1 = E_1 - mc^2$ is the laboratory kinetic energy of the incoming proton. Substitution of (6) in (5) leads to the threshold condition

$$T_1 \geq \overline{T}_1 = \left(2\mu + \frac{\mu^2}{2m}\right)c^2 \cong 280 \text{ MeV} \tag{7}$$

This prediction has received exact experimental confirmation: below $\overline{T}_1$ reaction (1) does not take place; above it is easily observed.

This last result can easily be generalized to the case of two particles with masses m_1 and m_2, in general different, that collide and produce in the final state a set of particles having a total rest mass M. If, for example, in the final state there

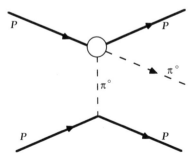

FIGURE 4.2. One pion exchange contribution to the process of π^0 production in proton–proton collisions.

TABLE 4.1. Laboratory Kinetic Energy Threshold for Some
Inelastic Processes

Reaction	Energetic threshold (MeV)
$P + P \to P + P + \pi^0$	280
$P + P \to P + \Lambda + K^+$	1585
$P + P \to P + \Sigma^+ + K^0$	1790
$P + P \to P + \Xi^- + K^+ + K^+$	3743
$\pi^+ + P \to \Lambda + K^+ + \pi^+$	1013
$\pi^- + P \to \Sigma^- + K^+$	904
$K^- + P \to \Xi^- + K^+$	662
$K^- + P \to \Omega^- + K^+ + K^+ + \pi^-$	3086

are three particles with rest masses m_1, m_2, and m_3, then $M = m_1 + m_2 + m_3$. In all cases the threshold for the inelastic process is

$$T_1 \geq \overline{T}_1 = \frac{M^2 - (m_1 + m_2)^2}{2m_2} c^2 \tag{8}$$

Using the known particle masses,[1] from Eq. (8) it is easy to calculate the thresholds of the reactions in Table 4.1.

The systematic agreement of these predictions with the empirical observations forms the basis of a physicist's faith in the conservation laws of energy and momentum. Furthermore, every experiment on elastic and inelastic collisions and every spontaneous disintegration confirm the validity of these laws, especially when the measurements imply an "overdetermined" reconstruction of the kinematical variables of the final state. This is the case when the particles taking part in a reaction are all electrically charged and, therefore, are all ionizing and measurable. An example is the last reaction of Table 4.1, in which all components of the final momenta are measured (they are $3 \times 4 = 12$) and their sums are compared with the momentum components of the incoming kaon.

4.1.2. Families of Particles

A dozen different particles appear in the eight examples of inelastic reactions of Table 4.1. These are subatomic systems endowed with the dual nature (wave + particle) characteristic of quantum systems, so the name "particles" is misleading. Furthermore, their large number makes it very unlikely that they are elementary.

With respect to stability, elementary particles can be of three types:

1. *Stable systems*: proton, electron, photon, perhaps neutrinos. These particles are stable because rigorous conservation laws prevent them from disintegrating into a set of lighter systems. As an example, consider

that the decay $P \rightarrow \pi^+ + \pi^+ + e^-$ would conserve energy, momentum, angular momentum, and electric charge; it never takes place because it would violate the baryon number conservation law.

2. *Metastable systems* are muons, pions, neutron, hyperons, and many others. They disintegrate spontaneously into lighter systems through the weak interaction (π^0 and Λ^0 through the electromagnetic interaction) with low probabilities and correspondingly large lifetimes (large with respect to the typical time of strong interactions: 10^{-23} s).

3. *Unstable systems*, for example the resonance ρ of two pions, the ϕ, which is very important for the EPR paradox, and the numerous resonances of the pion–nucleon type. These systems disintegrate immediately after being formed by a strong interaction, with lifetimes typically about 10^{-23} s.

Stable and metastable particles are usually classified in four different groups: *leptons*, *fundamental vector bosons*, *mesons*, and *baryons*. To each of these groups is devoted one of the tables of this section. The review of particle properties that we can afford to make in this section will necessarily be very partial in several respects: much experimental and conceptual evidence will therefore be sacrificed to simplicity.

In elementary particle physics the most important laws are the conservation laws. The available energy–momentum spontaneously reorganizes itself into one of the allowed forms of nature, as soon as the metastable or unstable form in which it was aggregated disintegrates. The subnuclear world is thus dominated by rigorous conservation laws, by some approximate conservation laws, and by a continuous reorganization of whatever exists. The property of existing, i.e., of being irreducibly real, must clearly be attributed to what is conserved in all reactions, and much less to what can disappear. From such a point of view electric charge is much more real than the muon, to give an example, and momentum is more real than velocity. The proton itself is less real than its baryon number: in the second of the inelastic reactions of which we calculated the threshold, for example, a proton disappears and is replaced by a lambda hyperon, which is different from the proton but also endowed with baryon number. One should not, however, go too far in denying the reality of what is not rigorously conserved; after all, we ourselves are not. In the following the four known families of particles are briefly reviewed.

4.1.2.1. Leptons

There exist three pairs of leptons: the electron, the muon, and the tau, each with a corresponding neutrino. All have spin $\frac{1}{2}$. The electron is by far the lightest charged lepton, followed by the muon. Neutrinos could all be massless, but for

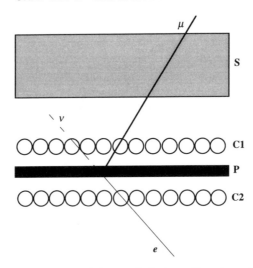

FIGURE 4.3. Apparatus used in the late thirties to study the decay of cosmic rays "mesotrons" (early name of muons). The large shield S was used to stop incoming electrons and photons. The circles represent Geiger counters, and P is a plate in which some mesotrons stopped and decayed at rest producing an electron (*e*) and a neutrino (ν). The electronics counted events where the pulse produced by the electron crossing the lower set of counters (C2) came between 1.5 and 20 μs after the pulse produced by the mesotron crossing the upper set C1. The correct identification of the muon as a kind of heavy electron is due to Pancini, Piccioni, and Conversi (1945).

two of them the experimental upper limit of the rest mass is still rather high. It is certain, however, that the three neutrinos differ physically. Table 4.2[1] (as in the following ones[1]) has four columns. The first contains symbols of the different leptons, the second their masses in units of MeV/c^2, the third the lifetimes in seconds, and the fourth the observed decays and their frequencies. For example $(0, 01)$ means that a decay takes place in 1% of all cases. If a particle can decay in different ways, it is believed that the choice between the various channels available is made at random at the time of disintegration. Why a particular tau should choose to decay into $e^- \bar{v}_e v_\tau$, another into $\pi^- \pi^0 v_\tau$, and so forth, remains to be understood.

A third, very rare, decay of the muon has been observed, but Table 4.2 (as all the following tables) lists only disintegrations having a probability not less than 1%. In the case of tau particles, which are much heavier, more than 20 different decays have been observed. Table 4.2 lists only a few of them (covering 66% of the total probability). Antileptons also exist, all different from the corresponding leptons. The positron (antielectron), the antimu, and the antitau have positive charge, but the same mass, the same spin, the same lifetime, and symmetrical decays of the electron, muon, and tau, respectively. Charged leptons and antileptons take part in weak and in electromagnetic interactions (but not in

TABLE 4.2. Leptons

Symbol	Mass (MeV/c^2)	Lifetime (s)	Decays (probabilities)
v_e	$<17 \times 10^{-6}$	stable	
v_μ	<0.27	?	
v_τ	<35	?	
e^-	0.511	stable	
μ^-	105	2.20×10^{-6}	$e^- \bar{v}_e v_\mu$ (0.99); $e^- \bar{v}_e v_\mu \gamma$ (0.01)
τ^-	1784	3.03×10^{-13}	$e^- \bar{v}_e v_\tau$ (0.18); $\mu^- \bar{v}_\mu v_\tau$ (0.18); $\pi^- \pi^0 v_\tau$ (0.23); $\pi^- \pi^+ \pi^- v_\tau$ (0.07); etc.

strong ones), neutrinos and antineutrinos only in weak interactions (unless they have hitherto unobserved magnetic moments).

4.1.2.2. Fundamental Vector Bosons

Fundamental vector bosons are spin-1 particles. The photon (γ) is the quantum of the electromagnetic field, and $W^\pm$ and Z are the quanta of the weak field. The classification of γ, $W^\pm$, and Z in the family of fundamental vector bosons arises from a theoretical prejudice (probably well founded) concerning the electroweak unification. It is remarkable that the existence of the heavy bosons of this family was first predicted theoretically, then observed experimentally more or less at the predicted values of the masses. At first sight $W^\pm$ and Z would appear to be very different from the photon, owing to their very large masses and their very short lifetimes (about 10 times smaller than the time required for light to cross a proton!). The decay channels of $W^\pm$ and Z are numerous: in Table 4.3 only the three simplest ones are reported, both for Z and W^+: anyway those of W^- are charge symmetrical to the reported ones. The γ and the Z coincide with their antiparticles, W^+ and W^- are each other's antiparticle.

4.1.2.3. Metastable Mesons

In Table 4.4 the metastable mesons with mass less than 2000 MeV/c^2 are listed. These particles all have spin 0 and negative parity. There are three π

TABLE 4.3. Fundamental Vector Bosons

Symbol	Mass (MeV/c^2)	Lifetime (s)	Decays (probabilities)
γ	$<3 \times 10^{-33}$	stable	
$W^\pm$	80600	10^{-24}	$e^+ v_e$ (0.10); $\mu^+ v_\mu$ (0.10); $\tau^+ v_\tau$ (0.10); etc.
Z	91160	10^{-24}	$e^+ e^-$ (0.03); $\mu^+ \mu^-$ (0.03); $\tau^+ \tau^-$ (0.03); etc.

TABLE 4.4. Metastable Mesons with mass less than $2000\,\text{MeV}/c^2$

Symbol	Mass (MeV/c^2)	Lifetime (s)	Decays (probabilities)
π^0	135	8.4×10^{-17}	2γ (0.99); $e^+e^-\gamma$ (0.01).
$\pi^\pm$	140	2.6×10^{-8}	$\mu^+\nu_\mu$ (1.00).
η^0	549	5.6×10^{-19}	2γ (0.39); $\pi^+\pi^-\pi^0$ (0.24); $3\pi^0$ (0.32); $\pi^+\pi^-\gamma$ (0.05).
$K^\pm$	494	1.2×10^{-8}	$\mu^+\nu_\mu$ (0.63); $\pi^0\mu^+\nu_\mu$ (0.03); $\pi^0 e^+\nu_e$ (0.05); $\pi^+\pi^0$ (0.21); $\pi^+\pi^0\pi^0$ (0.02); $\pi^+\pi^+\pi^-$ (0.06).
K_S^0	498	0.9×10^{-10}	$\pi^+\pi^-$ (0.69); $\pi^0\pi^0$ (0.31).
K_L^0	498	5.2×10^{-8}	$\pi^0\pi^0\pi^0$ (0.22); $\pi^+\pi^-\pi^0$ (0.12); $\pi^\pm\mu^\mp\nu$ (0.27); $\pi^\pm e^\mp\nu$ (0.39).
$D^\pm$	1869	1.1×10^{-12}	$\bar{K}^0\pi^+$ (0.03); $\bar{K}^0\pi^+\pi^0$ (0.08); $K^-\pi^+\pi^+$ (0.08); etc.
D^0	1864	4.2×10^{-13}	K^+K^- (0.05); $K^-\pi^+\pi^0$ (0.12); $K^-\pi^+\pi^+\pi^-$ (0.08); etc.
$D_S^\pm$	1969	4.5×10^{-13}	$\bar{K}^0K^+$ (0.03); $\pi^+\pi^+\pi^-$ (0.01): $\phi\pi^+$ (0.03); $\phi\pi^+\pi^0$ (0.06); $\phi\pi^+\pi^+\pi^-$ (0.01); etc.

mesons (pions, having charge $+e$, 0, $-e$), the neutral η^0, the four $K^\pm$, K_S^0, and K_L^0 (where $S=$ "short" and $L=$ "long" refer to the lifetimes) called "strange," the three D mesons called "charmed" (charge $+e$, 0, $-e$) having strangeness 0, and the two $D_S^\pm$ mesons both with charm and strangeness. The mesons of Table 4.4 are grouped in families of "isotopic spin" (also called isospin), in which the electric charge is the variable physical quantity. Pions have isospin 1 (three charge states), eta has isospin 0 (only one state), the K's are considered doublets (doublet of particles, K^+ and K^0 with strangeness $+1$, and doublet of antiparticles, K^- and $\bar{K}^0$ with strangeness -1); they therefore have isospin $\frac{1}{2}$. Notice that K_S^0 and K_L^0 are superpositions of K^0 and $\bar{K}^0$, in the quantum-mechanical sense. D^+ and D^0 also form a doublet (isospin $\frac{1}{2}$); D^- and $\bar{D}^0$ are the doublet of the antiparticles. Finally D_S^+ is a singlet (isospin 0) having D_S^- as antiparticle.

For charged particles only the decays of the positive one are reported: the negative case is symmetrical. The interaction determining the decays of π^0 and η^0 is the electromagnetic one: this explains the relatively short lifetimes. In all other cases one is dealing with weak interactions. The family of bosons also includes antiparticles: π^0 and η^0 are their own antiparticles, π^+ has π^- as its antiparticle, and so on.

4.1.2.4. Metastable Baryons

In Table 4.5 stable and metastable baryons whose mass is less than $2300\,\text{MeV}/c^2$ are listed. All such particles have spin $\frac{1}{2}$, with the possible

TABLE 4.5. Stable and Metastable Baryons with mass less than 2300 MeV/c^2

Symbol	Mass (MeV/c^2)	Lifetime (s)	Decays (probabilities)
P	938	stable	
N	940	890	$Pe^-\bar{\nu}_e$ (1.00).
Λ	1116	2.63×10^{-10}	$P\pi^-$ (0.64); $N\pi^0$ (0.36).
Σ^+	1189	0.80×10^{-10}	$P\pi^0$ (0.52); $N\pi^+$ (0.48).
Σ^0	1193	7.4×10^{-20}	$\Lambda\gamma$ (1.00).
Σ^-	1197	1.48×10^{-10}	$N\pi^-$ (1.00).
Ξ^0	1315	2.90×10^{-10}	$\Lambda\pi^0$ (1.00).
Ξ^-	1321	1.64×10^{-10}	$\Lambda\pi^-$ (1.00).
Ω^-	1672	0.82×10^{-10}	ΛK^- (0.68); $\Pi^0\pi^-$ (0.24); $\Pi^-\pi^0$ (0.08).
Λ_C^+	2285	1.91×10^{-13}	$PK^-\pi^+$ (0.03); $P\bar{K}^0\pi^+\pi^-$ (0.08); etc.

exception of the Ω, which could have spin $\frac{3}{2}$. There are two nucleons (charge $+e$, 0), one Λ (uncharged, strangeness -1), three strange hyperons Σ (charge $+e$, 0, $-e$, also with strangeness -1), two hyperons Ξ (charge 0, $-e$ and strangeness -2), one hyperon Ω (charge $-e$ and strangeness -3), and the "charmed" baryon Λ_c^+. Baryons are also grouped in families of isotopic spin, in which, again, electric charge is the variable physical quantity. Nucleons have isospin $\frac{1}{2}$ (two charge states), Λ has isospin 0 (only one state), the Σ's form a triplet (isospin 1), the Ξ's a doublet (isospin $\frac{1}{2}$), and the Ω a singlet (isospin zero). The Λ_c^+ is also a singlet (isospin 0).

Only the metastable baryons whose mass is less than 2300 MeV/c^2 have been included in Table 4.5. The proton is absolutely stable, or at least its lifetime is above 10^{31} years, that is, 10^{21} lives of the universe (with the Big Bang conjecture the universe is calculated to be about 15 billion years old). When a baryon decays there is always another baryon in the final state. This secondary baryon can, in turn, be metastable, but at the end of all possible disintegrations one is certain to find a proton. Many *conceivable* reactions, allowed by the other conservation laws (energy, momentum, angular momentum, electric charge), but which do not include a baryon in the final state, never occur. This empirical law is formulated as "baryon number conservation" and is obtained by attributing to each baryon a baryon number $+1$. To the corresponding antiparticles (antiproton, antineutron, antilambda, ...) the baryon number -1 is attributed. All baryons share with the proton this mysterious and indestructible property. Leptons, vector bosons, and mesons instead have baryon number 0.

In strong interactions strangeness is always conserved. The reader can check this for the eight inelastic reactions of Table 4.1: strangeness is 0, for example, in the initial and final states of the reaction $P + P \rightarrow P + \Sigma^+ + K^0$. In weak decays, on the other hand, strangeness is not conserved: a K^+, for instance, can decay into $\pi^+\pi^0$, final state with strangeness 0.

4.1.3. Nuclear Physics Experiments on the EPR Paradox

The first experiments on the EPR paradox were performed during the early seventies in low-energy nuclear physics, using gamma rays from electron–positron annihilation and the elastic scattering of two protons. Soon it was realized, however, that such experiments could not distinguish clearly between local realism and quantum theory, essentially because no known instrumentation can measure dichotomic observables of gamma rays and protons directly.

In the case of e^+e^- annihilation at rest, the resulting two photons have equal and opposite momenta. In 1950, Yang[2] showed that a photon polarization correlation arises as a consequence of invariance under rotations and reflection of the axes (parity), and that the appropriate state vector is

$$|\psi\rangle = \frac{1}{\sqrt{2}}\{|x_1\rangle|y_2\rangle - |y_1\rangle|x_2\rangle\} \tag{9}$$

where $|x_1\rangle$, $|y_1\rangle$ ($|x_2\rangle$, $|y_2\rangle$) are the linear polarization states of the first (second) photon along the x and y axes, respectively. As seen in the first chapter, eight years later Bohm and Aharonov used the experimental confirmation by Wu and Shaknov of the validity of such a description to rule out the Furry hypothesis.

In the experiment performed by Kasday, Ullman, and Wu[3] positrons from a ^{64}Cu source were annihilated in a thin layer of material surrounding the source. The annihilation γ rays were emitted in all directions, but the two vertical opposite directions were selected by a lead collimator before hitting two conical Compton scatterers (Fig. 4.4).

The scattered photons entered two scintillation detectors, covered with absorbers with a slit, such as to select photons at azimuthal angles ϕ_1 and ϕ_2. The three authors measured a quantity defined by

$$R(\phi_1, \phi_2) = \frac{N}{N_{ss}} \Bigg/ \frac{N_1}{N_{ss}}\frac{N_2}{N_{ss}}$$

where N is the number of times the two photons scatter and are both detected, N_1 the number of times the two photons scatter and only photon 1 is detected, N_2 the number of times the two photons scatter and only photon 2 is detected, and N_{ss} the number of times the two photons scatter.

Theoretically in all cases an expression such as

$$R(\phi_1, \phi_2) = A - B\cos 2(\phi_1 - \phi_2)$$

was expected, but different theories predicted different values of A and B, namely

$$A = 1, \quad B = B_0 \qquad \text{quantum mechanics}$$
$$A = 1, \quad B = B_0/\sqrt{2} \quad \text{upper limit of Bell's inequality}$$
$$A = 1, \quad B = B_0/2 \qquad \text{Furry's hypothesis}$$

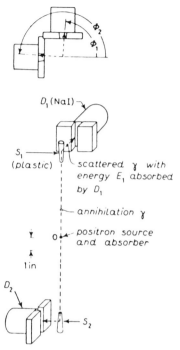

FIGURE 4.4. Schematic view of the apparatus of Kasday *et al.*[3] for measuring polarization correlations of e^+e^- annihilation photons. The positron source and the absorber are located near O. The photons travel in opposite directions and strike the Compton scatterers S1 and S2. The counters D1 and D2 detect the scattered γ rays. Φ_1 and Φ_2 are the azimuthal angles of the scattered photons.

The experimental results were only compatible with the quantum-mechanical values. However, Kasday himself pointed out in Varenna[3] that Bell's limit was not very meaningful for his experiment, because local hidden-variable models could easily explain the obtained experimental results.

Another experiment using e^+e^- annihilation was performed in Catania by Faraci, Gutkowski, Notarrigo, and Pennisi[4]. Their source was a ^{22}Na positron emitter enclosed in a Plexiglass container acting as an annihilator. Two plastic scintillators acted as Compton scatterers. Their results disagreed significantly with quantum mechanics and agreed with Bell's upper limit (Fig. 4.5). Furthermore, a surprising "distance effect" was found: the photon polarization correlation tended to decrease with separation and seemed to suggest a slow transition toward a mixture of factorizable states.

The puzzling result of this experiment[4] stimulated further investigations with e^+e^- annihilations, e.g., by Wilson, Lowe, and Butt[5] with a ^{64}Cu source of pairs of annihilation quanta. These authors wondered whether the correlation

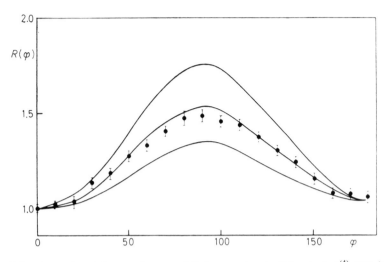

FIGURE 4.5. Results of the two-photon annihilation experiment of Faraci *et al.*[4] $R(\varphi)$ is the normalized angular correlation function dependent on the relative azimuthal angle φ. The upper curve is the quantum-mechanical prediction. The intermediate curve is the largest correlation function allowed by Bell's inequality (assumed applicable). The lower curve is the prediction of the Furry model. Systematic errors of normalization are possible in this and in the other experiments on photons from e^+e^- annihilation.

between the planes of polarization changed with source–scatterer separation. The source was usually placed symmetrically between the two scatterers, but measurements were also made for substantial differences in the separations. No significant change in the correlation was observed over separations of up to 2.5 m and over differences in separation of about 1 m (Fig. 4.6). A similar result, using a ^{22}Na positron emitter, was found a year later by Bruno, D'Agostino, and Maroni,[6] who, again, could find no evidence of a decrease in correlation as the distance between source and detectors increased.

Bertolini, Diana, and Scotti[7] in 1981 used ^{64}Cu as a source of annihilation photons. Their experiment differed from previous ones of the same type by the use of semiconductor detectors which allowed them to determine accurately the energies deposited by the photons in the scatterer and in the absorber of the Compton polarimeter (Fig. 4.7). Good agreement with the predictions of quantum mechanics was found after the experimental results were corrected for several secondary effects: multiple scattering, geometry of scatterers and detectors, etc.

Lamehi-Rachti and Mittig[8] carried out what has remained a unique experiment of its kind, using low-energy proton–proton scattering. During the collision between the two protons the interaction may be dominantly a "singlet" state scattering process, with the two spins of the protons antiparallel. In practice such a spin correlation experiment can be carried out by means of a twofold

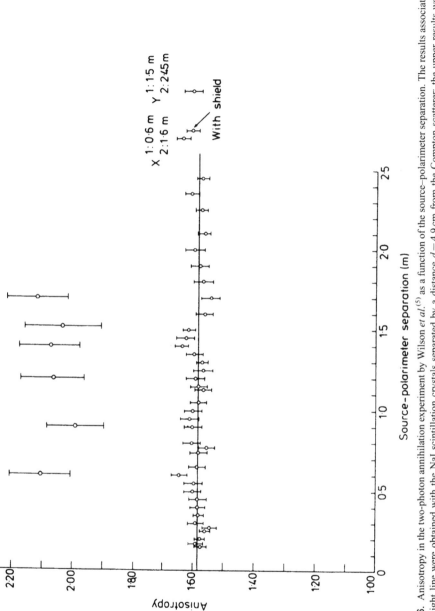

FIGURE 4.6. Anisotropy in the two-photon annihilation experiment by Wilson *et al.*[5] as a function of the source–polarimeter separation. The results associated with the straight line were obtained with the NaI scintillation crystals separated by a distance $d = 4.9$ cm from the Compton scatterer; the upper results were obtained with $d = 11$ cm. For both cases the positron source was placed symmetrically between the polarimeters, while for the results marked X and Y the source was positioned asymmetrically, the separations of the two polarimeters being indicated.

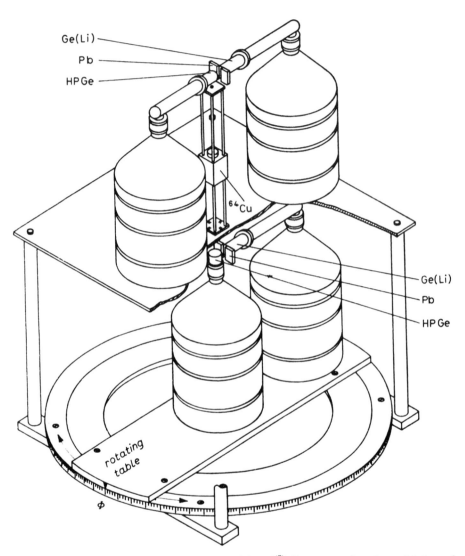

FIGURE 4.7. Apparatus of the experiment of Bertolini *et al*[7]. Gamma rays from the annihilations of positrons emitted by a ^{64}Cu source are 90° scattered by two germanium targets toward two semiconductor detectors. The distance between the scatterers (≈ 84 cm) was larger than the longitudinal coherence length of γ rays for positron annihilation in copper.

double-scattering experiment, as indicated in Fig. 4.8. A beam of 13.2-MeV protons hits a hydrogen target. After scattering at $\theta_{\text{lab}} = 45°$ each proton strikes a carbon foil and is scattered again. Four detectors for the doubly scattered protons measure the coincidence count rates $N_{LL}, N_{RR}, N_{LR}, N_{RL}$; where N_{LL} is the number of coincidences between the left counters L_1 and L_2, etc. In this experiment about 10^4 coincidences were analyzed. The detectors of one of the two carbon analyzers are in the scattering plane, while those of the second left–right scattering analyzer are rotated by an angle θ around the axis defined by protons entering the second analyzer. Lamehi-Rachti and Mittig defined the correlation function

$$P_{\text{meas}}(\theta) = \frac{N_{LL} + N_{RR} - N_{LR} - N_{RL}}{N_{LL} + N_{RR} + N_{LR} + N_{RL}} \tag{10}$$

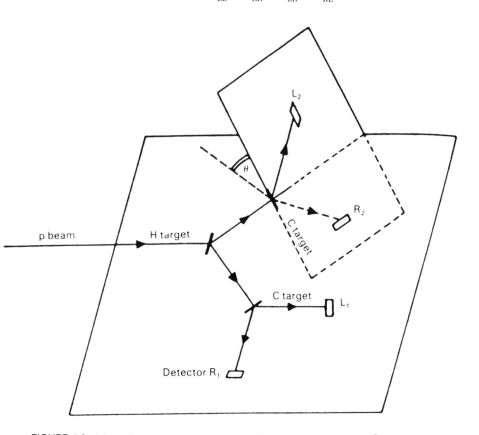

FIGURE 4.8. Schematic experimental arrangement of Lamehi-Rachti and Mittig[8] for the measurement of the spin correlation using two proton–carbon scatterings following the primary elastic scattering of the incoming proton in the H target.

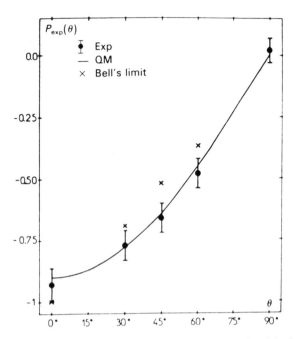

FIGURE 4.9. Experimental results for the spin correlation function found in the proton–proton double-scattering experiment carried out by Lamehi-Rachti and Mittig.[8] The results are compared with the limits of Bell's inequality (×) and quantum mechanics (QM).

The authors used an analogy with a double Stern–Gerlach experiment (not performed) for which the quantum-mechanical correlation function is predicted to be

$$P_{QM}(\theta) = -C_{nn} \cos \theta$$

Here C_{nn} is the Wolfenstein parameter, different from unity because of a contribution of the triplet state to proton–proton scattering and known at this energy to assume the value $C_{nn} = -0.950 \pm 0.015$.

The authors were well aware that their results could not be compared directly with Bell's inequality. Further assumptions were necessary, and they made the following:

1. It is, in principle, possible to construct a perfect apparatus.
2. The outcomes of measurements are not affected by the fact that the device used did not fulfil the conditions of spacelike separation.
3. The analyzing power and the transmission of the measuring apparatus can be considered intrinsic constants of the apparatus.

4. An upper limit for the contribution of triplet scattering can be obtained from the scattering of polarized protons on polarized protons.

Once these assumptions were made the authors could conclude that "the results are in good agreement with quantum mechanics and in disagreement with the inequality of Bell." Of course, the validity of the above assumptions, especially the third one, is highly questionable.

4.2. REALISM AND THE NEUTRAL KAON SYSTEM

In this section the compatibility between local realism and the physics of single kaons will be demonstrated. A physical model that reproduces the quantum-mechanical predictions of "strangeness oscillations" will be described. The irrelevance of Heisenberg's interpretation of the uncertainty relations and of Bohr's complementarity principle will thus once more be demonstrated. The most general local realistic treatment of single-kaon physics will be developed as well and expressed by means of upper and lower bounds on the relevant probabilities.

4.2.1. Neutral Kaons in Quantum Theory

It is assumed here that charge conjugation–parity is conserved: the operator CP commutes with the Hamiltonian. Otherwise the following is the standard phenomenological approach to the quantum-mechanical theory of neutral kaons. The CP eigenvalues are ± 1 and the eigenvectors are

$$CP|K_S\rangle = +|K_S\rangle \quad \text{and} \quad CP|K_L\rangle = -|K_L\rangle \tag{11}$$

Thus, the small effect of CP nonconservation is neglected in (11), as in most of this chapter, and the $CP = \pm 1$ eigenstates are identified with the short and long kaon, respectively.

The Hamiltonian H is regarded as being non-Hermitian to obtain an exponential decrease of the amplitudes for undecayed kaons. Thus, H satisfies the equations

$$H|K_S\rangle = \alpha_S|K_S\rangle \quad \text{and} \quad H|K_L\rangle = \alpha_L|K_L\rangle \tag{12}$$

where

$$\alpha_S = m_S - \frac{i}{2}\gamma_S, \qquad \alpha_L = m_L - \frac{i}{2}\gamma_L \tag{13}$$

In Eq. (13), γ_S and m_S (γ_L and m_L) denote the decay rate and mass, respectively, of the $S(L)$ meson. The useful numerical parameters in units $\hbar = c = 1$ are

$$\gamma_S = (1.121 \pm 0.002) \times 10^{10}\ s^{-1}, \qquad \gamma_L = (1.934 \pm 0.015) \times 10^{7}\ s^{-1}$$

$$\Delta m = m_L - m_S = (0.535 \pm 0.003) \times 10^{10}\ s^{-1}$$

Taking γ_S as inverse time unit they can be written instead as

$$\gamma_S = 1, \quad \gamma_L = \frac{1}{579.6}, \quad \Delta m = \frac{1}{2.10}$$

The time evolution operator is written in terms of the Hamiltonian H:

$$U(t) = \exp\{-iHt\} \tag{14}$$

where t is the particle proper time. Given that H commutes with CP, $U(t)$ also does and one has

$$\begin{aligned}
|K_S(t)\rangle &= U(t)|K_S(0)\rangle = e^{-\alpha_S t}|K_S(0)\rangle \\
|K_L(t)\rangle &= U(t)|K_L(0)\rangle = e^{-\alpha_L t}|K_L(0)\rangle
\end{aligned} \tag{15}$$

where, because of (13), the complex exponent gives rise to an exponential decrease of the wave function.

The second fundamental quantum number is strangeness, S. Its eigenvalues and eigenvectors satisfy

$$S|K^0\rangle = +|K^0\rangle \quad \text{and} \quad S|\overline{K}^0\rangle = -|\overline{K}^0\rangle \tag{16}$$

Strangeness does not commute with the weak Hamiltonian H. At time 0 the relations

$$\begin{aligned}
|K^0(0)\rangle &= \frac{1}{\sqrt{2}}[|K_S(0)\rangle + |K_L(0)\rangle] \\
|\overline{K}^0(0)\rangle &= \frac{1}{\sqrt{2}}[|K_S(0)\rangle - |K_L(0)\rangle]
\end{aligned} \tag{17}$$

between the S and CP eigenvectors hold. By applying the time evolution operator to Eq. (17), one sees that the strangeness eigenvectors evolve into

$$\begin{aligned}
|K^0(t)\rangle &= \frac{1}{\sqrt{2}}[e^{-\alpha_S t}|K_S(0)\rangle + e^{-\alpha_L t}|K_L(0)\rangle] \\
|\overline{K}^0(t)\rangle &= \frac{1}{\sqrt{2}}[e^{-\alpha_S t}|K_S(0)\rangle - e^{-\alpha_L t}|K_L(0)\rangle]
\end{aligned} \tag{18}$$

where t is the kaon proper time. By inverting (17) and substituting the result into (18), one easily obtains

$$|K^0(t)\rangle = \tfrac{1}{2}[e^{-\alpha_S t} + e^{-\alpha_L t}]|K^0(0)\rangle + \tfrac{1}{2}[e^{-\alpha_S t} - e^{-\alpha_L t}]|\overline{K}^0(0)\rangle$$
$$|\overline{K}^0(t)\rangle = \tfrac{1}{2}[e^{-\alpha_S t} - e^{-\alpha_L t}]|K^0(0)\rangle + \tfrac{1}{2}[e^{-\alpha_S t} + e^{-\alpha_L t}]|\overline{K}^0(0)\rangle$$

(19)

Therefore, a kaon born with fixed strangeness evolves with time into a superposition of the two strangeness eigenstates. For example, the probability that a $|K^0(t)\rangle$ state (born as an $S = +1$ eigenstate) is observed to behave as an $S = -1$ neutral kaon is given by

$$\tfrac{1}{4}|e^{-\alpha_S t} - e^{-\alpha_L t}|^2 = \tfrac{1}{4}[e^{-\gamma_S t} + e^{-\gamma_L t} - 2e^{-\gamma_S t/2}e^{-\gamma_L t/2}\cos \Delta m\, t]$$

The interference term gives rise to the well-known phenomenon of "strangeness oscillation." By inverting (17) and applying the time evolution operator, one obtains

$$|K_S(t)\rangle = e^{-\alpha_S t}\frac{1}{\sqrt{2}}[|K^0(0)\rangle + |\overline{K}^0(0)\rangle]$$

(20)

$$|K_L(t)\rangle = e^{-\alpha_L t}\frac{1}{\sqrt{2}}[|K^0(0)\rangle - |\overline{K}^0(0)\rangle]$$

For single kaons there is nothing paradoxical about (19)–(20). Indeed, in the next two sections it will be shown that local realistic models exist that reproduce all the empirical consequences of (19)–(20).

Neutral kaon pairs are generated in the decay of ϕ mesons at rest produced in electron–positron collisions:

$$e^+ + e^- \rightarrow \phi \rightarrow K^0 + \overline{K}^0$$

Given the quantum numbers of the ϕ and the usual conservation laws of angular momentum J, parity P, and charge conjugation C in its (strong) decay, the final neutral kaons are described in quantum mechanics by the $J^{PC} = 1^{--}$ state vector

$$|\psi\rangle = \frac{1}{\sqrt{2}}\{|K^0\rangle_a|\overline{K}^0\rangle_b - |\overline{K}^0\rangle_a|K^0\rangle_b\} = \frac{1}{\sqrt{2}}\{|K_S\rangle_a|K_L\rangle_b - |K_L\rangle_a|K_S\rangle_b\} \quad (21)$$

where a (left) and b (right) denote the opposite directions of motion of the kaons. The small effect of CP nonconservation is neglected also in (21), as in most of this chapter. The time evolution operator of vector (21) is the product of the evolution operators for the individual kaons; hence, at times t_a and t_b one has

$$|\psi(t_a, t_b)\rangle = \frac{1}{\sqrt{2}}\{|K_S\rangle_a|K_L\rangle_b \exp(-\alpha_S t_a - \alpha_L t_b) - |K_L\rangle_a|K_S\rangle_b \exp(-\alpha_L t_a - \alpha_S t_b)\}$$

(22)

Time 0 will be understood where no time is specified. The difference between the two exponentials in (22) generates $K^0 K^0$ and $\overline{K}^0 \overline{K}^0$ components. The probability of $\overline{K}^0 \overline{K}^0$ observation at times t_a and t_b is

$$P_{QM}[\overline{K}(t_a); \overline{K}(t_b)] = \tfrac{1}{8}\{e^{-\gamma_S t_a - \gamma_L t_b} + e^{-\gamma_L t_a - \gamma_S t_b} - 2e^{-\gamma(t_a + t_b)/2} \cos \Delta m(t_a - t_b)\}$$
(23)

where $\gamma = \gamma_S + \gamma_L$ and $\Delta m = m_L - m_S$ is the $K_L - K_S$ mass difference. The right side of Eq. (23) vanishes for $t_a = t_b$, as it must. In a real experiment the detection of $\overline{K}^0$'s can be achieved either via hyperon production in two suitably placed targets or via $\Delta S = \Delta Q$ semileptonic decays at appropriate distances from the ϕ decay region.

4.2.2. Reinterpretation of Quantum Probabilities

Local realism in the case of neutral kaons consists of the following three assumptions:

1. If, without in any way disturbing a kaon, one can predict with certainty the value of a physical quantity of that kaon, then there exists an element of reality corresponding to this physical quantity (EPR *reality criterion*).
2. If two kaons are very far apart, an element of reality belonging to one of them cannot be created by a measurement performed on the other (*locality*).
3. If at a given time t a kaon has an element of reality, the latter cannot be created by measurements on the same kaon performed at time t', if $t' > t$ (*no retroactive causality*).

It is very useful to consider those predictions of (22) to which the EPR reality criterion can be applied. These are the strict anticorrelations in strangeness S and in CP. If they are assumed to be exact the following conclusions hold:[9]

(i) Each kaon of every pair has an element of reality λ_1 that determines a well-defined value of CP ($\lambda_1 = \pm 1$ corresponds to $CP = \pm 1$, respectively).

(ii) Each kaon of every pair has an element of reality λ_2 that determines a well-defined strangeness S ($\lambda_2 \pm 2$ corresponds to $S = \pm 1$, respectively).

Furthermore, λ_1 is a stable property, while λ_2 undergoes sudden jumps from $S = +1$ to $S = -1$, and *vice versa*, which are simultaneous for the two kaons of every pair but occur at random times in a statistical ensemble of many pairs.

The application of local realism to the physical situation described quantum mechanically by (22) already represents a departure from quantum theory, at least formally: no quantum-mechanical state exists, in fact, describing a kaon as having simultaneously well-defined CP and S values. In the local realistic approach the observables S and CP, described quantum mechanically by two noncommuting operators, are simultaneously predetermined by elements of reality belonging to any given kaon. This is the standard treatment of "incompatible" observables in all "hidden variable" theories. The necessary codefinition of S and CP can be rigorously justified, as we have seen, by applying local realism to a kaon belonging to a pair. Nevertheless, it is natural to assume that all kaons have the same basic properties and to extend this codefinition to single kaons even when they do not belong to an EPR pair. The following developments depend on this natural extension.

One can easily reproduce the quantum-mechanical predictions for strangeness oscillations and decay of single K^0-mesons within the local realistic approach. Following the foregoing ideas, one introduces the four basic states of local realism:

$K_1 \equiv K_S$:   state with $S = +1$ and $CP = +1$ (short-living kaon)

$K_2 \equiv \overline{K}_S$:   state with $S = -1$ and $CP = +1$ (short-living antikaon)

$K_3 \equiv K_L$:   state with $S = +1$ and $CP = -1$ (long-living kaon)   (24)

$K_4 \equiv \overline{K}_L$:   state with $S = -1$ and $CP = -1$ (long-living antikaon)

Next the probabilities of their descriptions holding in a given physical situation are introduced:

$$p_i(t) = \text{probability of } K_i \text{ at proper time } t \qquad (i = 1, 2, 3, 4) \qquad (25)$$

The initial conditions depend on the problem considered. If one assumes, as an example, that initially only states with $S = +1$ are produced with equiprobable $CP = \pm 1$ components, one obtains the situation described in quantum theory by the $S = +1$ ket (17). Therefore,

$$p_1(0) = p_3(0) = \tfrac{1}{2}, \qquad p_2(0) = p_4(0) = 0 \qquad (26)$$

To agree with the experimentally well-established validity of the quantum-mechanical probabilities (which refer to well defined initial S without specifying CP), one must find local realistic models reproducing the following equalities for the chosen state:

$$p_1(t) + p_2(t) = |\langle K_S(0)|K(t)\rangle|^2 = \tfrac{1}{2} E_S$$
$$p_3(t) + p_4(t) = |\langle K_L(0)|K(t)\rangle|^2 = \tfrac{1}{2} E_L \qquad (27)$$

and

$$p_1(t) + p_3(t) = |\langle K(0)|K(t)\rangle|^2 = \tfrac{1}{4}[E_L + E_S + 2\sqrt{E_L E_S} \cos \Delta m\ t]$$

$$p_2(t) + p_4(t) = |\langle \overline{K}(0)|K(t)\rangle|^2 = \tfrac{1}{4}[E_L + E_S - 2\sqrt{E_L E_S} \cos \Delta m\ t] \tag{28}$$

Notice that (27) and (28) are compatible with (26). The sum of Eqs. (27) is equal to the sum of Eqs. (28). Hence, there are only three independent conditions for the four probabilities (25).

4.2.3. The Probabilities of Local Realism

It is easy to verify that (27) and (28) can be reproduced by writing[10]

$$p_1 = \tfrac{1}{2}E_S Q_+, \quad p_2 = \tfrac{1}{2}E_S Q_-, \quad p_3 = \tfrac{1}{2}E_L Q_+, \quad p_4 = \tfrac{1}{2}E_L Q_- \tag{29}$$

where

$$Q_\pm(t) \equiv \frac{1}{2}\left[1 + \frac{2\sqrt{E_L E_S}}{E_L + E_S} \cos \Delta m\ t\right] \tag{30}$$

These results can be rewritten in a physically more appealing way as

$$p_1(t) = \frac{1}{2}e^{-\gamma_S t}\frac{|\psi_L + \psi_S|^2}{2\{|\psi_L|^2 + |\psi_S|^2\}} \qquad p_2(t) = \frac{1}{2}e^{-\gamma_S t}\frac{|\psi_L - \psi_S|^2}{2\{|\psi_L|^2 + |\psi_S|^2\}}$$

$$p_3(t) = \frac{1}{2}e^{-\gamma_L t}\frac{|\psi_L + \psi_S|^2}{2\{|\psi_L|^2 + |\psi_S|^2\}} \qquad p_4(t) = \frac{1}{2}e^{-\gamma_L t}\frac{|\psi_L - \psi_S|^2}{2\{|\psi_L|^2 + |\psi_S|^2\}} \tag{31}$$

from which it is clear that all probabilities are positive, as they should be. In (31) the following "wave functions" were introduced:

$$\psi_L = \psi_0 \exp(-\gamma_L t/2)\exp(-im_L t), \qquad \psi_S = \psi_0 \exp(-\gamma_S t/2)\exp(-im_s t) \tag{32}$$

Notice that the simple equations (31) do not exist within standard quantum theory. They suggest a dualistic picture according to which all kaons are particles embedded in extended waves, which are in all cases superpositions of ψ_L and ψ_S, and evolve according to (32) with the same (unknown) initial value ψ_0.[11] The probabilities (31) find a complete physical interpretation within this model. For example, $p_1(t)$ can be understood as follows: the factor $\tfrac{1}{2}$ is the probability that the given kaon is born with $CP = +1$, in agreement with (26); the exponential factor is the probability that it remains undecayed at time t; and the final fraction is the probability that it has positive strangeness at time t. In this way the quantum-mechanical probabilities (27) and (28) are also given a completely new physical interpretation within the local realistic approach. The quantum ensembles are interpreted to be mixtures of other ensembles in which the basic states of local realism (24) apply. It should be stressed that Eqs. (31) do not give the most general probabilities within local realism. A complete generalization will be given later.

The probabilities (29) describe a particular mixture, that is, a statistical ensemble in which two of the basic states (24) are initially present with equal statistical weights, as shown by the initial conditions (26). The physical reinterpretation given in the previous section allows us, however, to extend very naturally our results to the case of "pure states" (only one kaonic state produced initially), which are needed in the EPR problem. Probabilities with two indices will be used, the second one specifying which of the four states (24) was initially present. Naturally there are four possibilities.

In a given situation the basic probabilities for the four states (24) at proper time t can be considered conditional on the initial presence of $K_1(0)$, $K_2(0)$, $K_3(0)$, or $K_4(0)$. Using the symbol $p_{ji}(t|0)$ to denote the probability of a kaon in state K_j at proper time t conditional on the same kaon having been in state K_i at proper time 0 ($j, i = 1, 2, 3, 4$) and remembering that in eight cases the probability is zero owing to CP conservation, one can write the remaining eight probabilities:

$$
\begin{aligned}
p_{11}(t|0) &= E_S(t)Q_+(t), & p_{21}(t|0) &= E_S(t)Q_-(t) \\
p_{12}(t|0) &= E_S(t)Q_-(t), & p_{22}(t|0) &= E_S(t)Q_+(t) \\
p_{33}(t|0) &= E_L(t)Q_+(t), & p_{43}(t|0) &= E_L(t)Q_-(t) \\
p_{34}(t|0) &= E_L(t)Q_-(t), & p_{44}(t|0) &= E_L(t)Q_+(t)
\end{aligned}
\tag{33}
$$

which satisfy obvious initial conditions, for example,

$$
p_{11}(0|0) = 1, \qquad p_{21}(0|0) = 0
$$

The $p_{ji}(t|0)$ appearing in (33) have a physical interpretation similar to that given for $p_1(t)$ at the end of the previous section, only relating to different initial conditions. In the case of $p_{11}(t|0)$ the given kaon is born with $CP = S = +1$, $E_S(t)$ is the probability that it has not yet decayed at time t, and $Q_+(t)$ is the probability of strangeness remaining positive at time t. The interpretation of all $p_{ji}(t|0)$ ($j, i = 1, 2, 3, 4$) in (33) and in the equations to follow is always similar.

The probabilities introduced through (33) will be said to constitute "the standard set." Of course these probabilities are to some extent arbitrary, but a complete generalization will be given later.

4.2.4. The Most General Set of Probabilities

In the next sections the shorter notation p_{ji} instead of $p_{ji}(t|0)$ will be used for conditional probabilities. One can check that all quantum-mechanical probabilities for single kaons are reproduced. For example, CP conservation is satisfied,

$$
\begin{aligned}
|\langle K_L(0)|K_S(t)\rangle|^2 &= 0 = \tfrac{1}{2}[p_{31} + p_{41} + p_{32} + p_{42}] \\
|\langle K_S(0)|K_L(t)\rangle|^2 &= 0 = \tfrac{1}{2}[p_{13} + p_{23} + p_{14} + p_{24}]
\end{aligned}
\tag{34}
$$

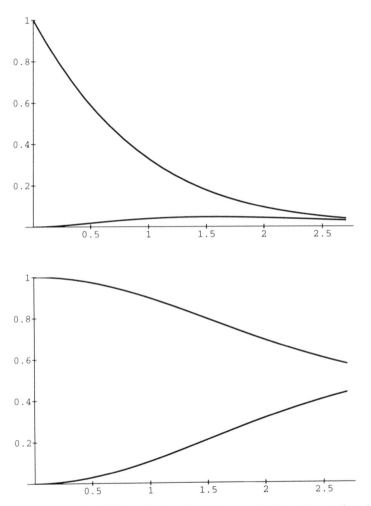

FIGURE 4.10. (Above) Probabilities of local realism for kaons with $CP = +1$: p_{11} and p_{22} (upper curve); p_{12} and p_{21} (lower curve). (Below) Probabilities of local realism for kaons with $CP = -1$: p_{33} and p_{44} (upper curve); p_{34} and p_{43} (lower curve). Proper time in abscissa (units γ_S^{-1}).

because all terms on the right sides of (34) vanish. It is easy to verify that all 14 conditions deducible from the state vectors (19)–(20) are satisfied by the probabilities (33). These make up the "standard probability matrix":

$$
\mathbf{M}_0 = \begin{vmatrix} E_S Q_+ & E_S Q_- & 0 & 0 \\ E_S Q_- & E_S Q_+ & 0 & 0 \\ 0 & 0 & E_L Q_+ & E_L Q_- \\ 0 & 0 & E_L Q_- & E_L Q_+ \end{vmatrix} \tag{35}
$$

where $p_{11}(t|0) = E_S(t)Q_+(t)$, $p_{12}(t|0) = E_S(t)Q_-(t)$, etc. It was shown in Ref. 11 that the most general probability matrix that satisfies all 16 quantum-mechanical conditions that can be extracted from the state vectors (18–20) is

$$\mathbf{M} = \begin{vmatrix} E_S Q_+ + \rho & E_S Q_- - \rho & 0 & 0 \\ E_S Q_- - \rho & E_S Q_+ + \rho & 0 & 0 \\ 0 & 0 & E_L Q_+ - \rho & E_L Q_- + \rho \\ 0 & 0 & E_L Q_- + \rho & E_L Q_+ - \rho \end{vmatrix} \tag{36}$$

It is ρ that makes this the most general formulation of local realism. Every column in (36) refers to a well-defined initial state, one of the four states (24). The sum of the elements of a column equals E_S [for $K_1(0)$ and $K_2(0)$] and E_L [for $K_3(0)$ and $K_4(0)$]. This accounts for population reduction due to spontaneous disintegration. Restrictions on ρ can be obtained by assuming every probability to be positive and less than 1. One gets

$$-E_S Q_+ \le \rho \le E_S Q_-, \qquad -E_L Q_- \le \rho \le E_L Q_+ \tag{37}$$

where the inequalities with $E_S(E_L)$ were obtained from the first and second (third and fourth) columns of $\mathbf{M}$. Conditions (37) must both be satisfied in any consistent local realistic theory. It was checked numerically that at all times one has $E_S(t)Q_-(t) \le E_L(t)Q_+(t)$. In other words, of the upper limits in (37) it is enough to consider

$$\rho(t) \le \rho^{\max}(t) = E_S(t)Q_-(t) \tag{38}$$

The other one is automatically satisfied. No simplification of this type exists for the lower limits of Eq. (37). In this case it was checked numerically that the equation

$$E_S(t)Q_+(t) = E_L(t)Q_-(t)$$

is satisfied only for $t = t_0 = 1.431\gamma_s^{-1}$. Below (above) t_0, $-E_L(t)Q_-(t)$ is larger (smaller) than $-E_S(t)Q_+(t)$. Therefore,

$$\rho(t) \ge \rho^{\min}(t) = \begin{cases} -E_L(t)Q_-(t) & \text{if } t < 1.431\gamma_s^{-1} \\ -E_S(t)Q_+(t) & \text{if } t > 1.431\gamma_s^{-1} \end{cases} \tag{39}$$

Equations (38) and (39) are necessary conditions that all local realistic theories of single-kaon physics have to satisfy. Notice that $Q_+(t)$ and $Q_-(t)$ as defined in (30) are always positive, $Q_-(0) = 0$ excepted. Therefore the upper bound (38) [lower bound (39)] is never negative (never positive).

By remaining in the frame of single-kaon physics it is not possible to say more. It can be shown, however, that the comparison of local realism with the nonparadoxical predictions of quantum theory for correlated kaon pairs leads necessarily to $\rho(t) = 0$, so the local realistic theory of single kaons becomes unique. All this will be seen in Section 4.4.

4.3. THE DEBATE ON THE EPR PARADOX FOR PAIRS OF PARTICLES

About a dozen proposals for the experimental study of the EPR paradox in particle physics are critically reviewed. Some of them seem to open new interesting possibilities, especially with new accelerators currently under construction (ϕ factories, B factories).

4.3.1. Forbidden Symmetrical Observations

As shown in Section 4.2.1, pairs of neutral kaons can be generated in the decay of ϕ mesons, (e.g., produced at rest in the laboratory in e^+–e^- collisions):

$$e^+ + e^- \rightarrow \phi \rightarrow K^0 + \overline{K}^0$$

The two final neutral kaons are described quantum mechanically by the $J^{PC} = 1^{--}$ state vector (22) for arbitrary proper times of the two kaons.

The probability of a double $\overline{K}^0$ observation at times t_a and t_b is given by (23) and can be written

$$P_{QM}[\overline{K}(t_a); \overline{K}(t_b)] = \tfrac{1}{8}[E_S(t_a)E_L(t_b) + E_L(t_a)E_S(t_b)$$
$$-2\sqrt{E_S(t_a)E_L(t_b)E_L(t_a)E_S(t_b)}\cos \Delta m(t_a - t_b)] \quad (40)$$

where

$$E_S = e^{-\gamma_S t}, \qquad E_L = e^{-\gamma_L t} \quad (41)$$

and $\Delta m = m_L - m_S$ is the $K_L - K_S$ mass difference.

As discussed in Chapter 1, Lipkin[12] showed that there is an important consequence of the state vector (22): the observation of a particular decay mode, e.g., $\pi^+\pi^-$, for one kaon implies that the other kaon cannot decay, *at the same proper time*, in this mode. The simultaneous decay of both kaons into two neutral pions is similarly forbidden, as are all equal decays. Table 4.6 lists the pairs of decays forbidden at equal proper times. One can show that the previous selection rule applies even if CP violation is taken into account, because CP violation

TABLE 4.6. Pairs of Decays Forbidden at Equal
Proper Times

$(K^0)_a \rightarrow \pi^+ + \pi^-$	$(K^0)_b \rightarrow \pi^+ + \pi^-$
$(K^0)_a \rightarrow \pi^0 + \pi^0$	$(K^0)_b \rightarrow \pi^0 + \pi^0$
$(K^0)_a \rightarrow \pi^+ + \pi^- + \pi^0$	$(K^0)_b \rightarrow \pi^+ + \pi^- + \pi^0$
$(K^0)_a \rightarrow \pi^+ + e^- + \bar{\nu}_e$	$(K^0)_b \rightarrow \pi^+ + e^- + \bar{\nu}_e$
$(K^0)_a \rightarrow \pi^- + e^+ + \nu_e$	$(K^0)_b \rightarrow \pi^- + e^+ + \nu_e$
$(K^0)_a \rightarrow \pi^+ + \mu^- + \bar{\nu}_\mu$	$(K^0)_b \rightarrow \pi^+ + \mu^- + \bar{\nu}_\mu$
$(K^0)_a \rightarrow \pi^- + \mu^+ + \nu_\mu$	$(K^0)_b \rightarrow \pi^- + \mu^+ + \nu_\mu$

modifies (22) only in the multiplying factor, while the term in braces remains the same.

The essential physical point of these phenomena can be summarized as follows: The observation of a particular decay mode at time t in the positive (a) direction imposes constraints on the kaon beam observed in coincidence in the negative (b) direction. These constraints force $|K\rangle_b$ to be a definite linear combination of $|K_S\rangle$ and $|K_L\rangle$. It is the very linear combination that at the same t cannot decay in that mode.

In a paper published in the DAϕNE Physics Handbook, Baldini–Celio and collaborators[13] pointed out another aspect of the EPR paradox for neutral kaons: Even a state vector with sharp CP anticorrelation like (22) could, in principle, give rise to two K_S mesons if a thin regenerator, which converts K_L into K_S, is introduced on one side of the experiment. Coherent regeneration cannot arise, however, for a spherical regenerator because the state vector (22) is invariant under that *simultaneous* transformation of the state vectors of the two kaons that represents coherent regeneration. The doughnut of the storage ring is considered by these authors to be an appropriate regenerator for the verification of this prediction.

4.3.2. Some Applications of Furry's Hypothesis

The "Furry hypothesis" consists of the spontaneous evolution of the state vector (21) into its components, e.g., as follows:

$$\begin{aligned} |\psi\rangle \Rightarrow |K_S\rangle_a |K_L\rangle_b \qquad &\text{in half of the cases} \\ |\psi\rangle \Rightarrow |K_L\rangle_a |K_S\rangle_b \qquad &\text{in the other half} \end{aligned} \qquad (42)$$

Six[14] viewed this as a possible evolution for correlated kaon pairs (taking place immediately after the decay of the ϕ meson). In 1991 Piccioni[15] also considered the Furry hypothesis for the same process. He pointed out that the available experimental evidence already showed that only $K_S K_L$ pairs were produced, no $K_S K_S$ or $K_L K_L$ pairs. To be consistent with this evidence, the Furry hypothesis had to be assumed to lead from the state vector (21) to the mixture (42) and not to a mixture of $|K^0 \overline{K}^0\rangle$ and $|\overline{K}^0 K^0\rangle$, which would also be possible in principle.

The main virtue of the mixture (42) is that it avoids the Einstein–Podolsky–Rosen paradox (it is compatible with local realism). A serious drawback is that it contradicts quantum theory even for the (nonparadoxical) Lipkin anticorrelations in strangeness. The predicted probability P_F for two $\overline{K}^0$ observations differs from the quantum-mechanical probability by the absence of the interference term in (40):

$$P_F[\overline{K}(t_a); \overline{K}(t_b)] = \tfrac{1}{8}[E_S(t_a)E_L(t_b) + E_L(t_a)E_S(t_b)] \qquad (43)$$

One easy way to rule out the Furry hypothesis is simply to measure the strangeness of the two kaons: sharp anticorrelations are predicted by (40), but should be totally absent if the Furry hypothesis (43) holds.

The problem of quantum nonseparability versus local realism for $B^0\overline{B}^0$ pairs was discussed by Datta and Home,[16] who considered the experimental evidence concerning the decay of the spin-1 Y(4s) meson into a pair of pseudoscalar mesons $B^0\overline{B}^0$. The quantum formalism used for this $B^0\overline{B}^0$ system is very similar to the one used for $K^0\overline{K}^0$ pairs, with the difference that the states of the B^0 meson analogous to K_S and K_L have the same decay width. Datta and Home deduced some consequences of the Furry hypothesis and showed that an empirically measurable quantity R should take the value $R_F = 1$, while the corresponding quantum-mechanical prediction was $R_{QM} \ll 1$. Preliminary experimental evidence was in favor of the latter prediction.

Stronger evidence against the validity in nature of Furry's hypothesis was recently found at the empirical level by Muller[17] and collaborators in Saclay. In this experiment $K^0\overline{K}^0$ pairs were produced by antiprotons annihilating in a gaseous hydrogen target. The $\bar{p} + p$ annihilations at rest are known to take place mainly in the $^3S^1$ state or in the $^3P^0$ state. In the former case the $K^0\overline{K}^0$ channel has the quantum numbers $J^{PC} = 1^{--}$ and the two kaons are represented by the wave function (21), while in the second case the quantum numbers are $J^{PC} = 0^{++}$ and the wave function at time of decay is

$$|\psi_0\rangle = \frac{1}{\sqrt{2}}\{|K^0\rangle_a|\overline{K}^0\rangle_b + |\overline{K}^0\rangle_a|K^0\rangle_b\} = \frac{1}{\sqrt{2}}\{|K_S\rangle_a|K_S\rangle_b - |K_L\rangle_a|K_L\rangle_b\} \quad (44)$$

Clearly the last state leads only to K_SK_S and K_LK_L pairs, at least in the approximation of CP conservation, and can be distinguished experimentally from the state leading to K_SK_L pairs represented by (22). The experimental ratio of the corresponding rates depends on the hydrogen pressure, and in a typical measurement carried out at the Low-Energy Antiproton Ring (LEAR) at CERN was found to be

$$\frac{R(K_SK_S)}{R(K_SK_L)} = 0.037 \pm 0.002$$

so the contamination of the "unwanted" $J^{PC} = 0^{++}$ channel is fairly small. This means that the prediction (40) for the double-time probability of observing the $K^0\overline{K}^0$ pair should be reasonably accurate. The same can be said for

$$P_{QM}[K(t_a); \overline{K}(t_b)] = \tfrac{1}{8}[E_S(t_a)E_L(t_b) + E_L(t_a)E_S(t_b)$$
$$+ 2\sqrt{E_S(t_a)E_L(t_b)E_L(t_a)E_S(t_b)} \cos \Delta m(t_a - t_b)] \quad (45)$$

which is also a consequence of the state vector (22). The quantum-mechanical probabilities (40) and (45) are proportional to the corresponding intensities:

$$I[K(t_a); \overline{K}(t_b)] = kP_{QM}[K(t_a); \overline{K}(t_b)], \qquad I[\overline{K}(t_a); \overline{K}(t_b)] = kP_{QM}[\overline{K}(t_a); \overline{K}(t_b)]$$

where k is a constant. Muller's experiment measured the asymmetry

$$\alpha[t_a, t_b] = \frac{I[K(t_a); \overline{K}(t_b)] - I[\overline{K}(t_a); \overline{K}(t_b)]}{I[K(t_a); \overline{K}(t_b)] + I[\overline{K}(t_a); \overline{K}(t_b)]}$$

which, from (40) and (45), is predicted to be

$$\alpha[t_a, t_b] = \frac{2\sqrt{E_S(t_a)E_L(t_b)E_L(t_a)E_S(t_b)} \cos \Delta m(t_a - t_b)}{E_S(t_a)E_L(t_b) + E_L(t_a)E_s(t_b)}$$

According to the Furry hypothesis the interference terms should be absent from (22) and (45), and the asymmetry should accordingly vanish. The kaon strangeness was measured via scattering in a copper target where only the negative strangeness state can lead to hyperon production. The result $\alpha = 0.84 \pm 0.17$, averaged over some time interval, was clearly different from zero and thus ruled out Furry's hypothesis.

What is particularly interesting in the experiment is that its natural development in the near future could, in principle, provide important information concerning the comparison between local realism and quantum mechanics, even before the ϕ factory accelerator starts to operate.

4.3.3. Other Proposed Forms of the Paradox

In 1986 Törnqvist[18] considered the reaction

$$e^+ + e^- \to J/\psi \to \Lambda + \overline{\Lambda} \tag{46}$$

followed by the decays $\Lambda \to \pi^- + p$ and $\overline{\Lambda} \to \pi^+ + \overline{p}$, and hoped to produce a violation of Bell's inequality. He calculated the quantum-mechanical rate of the overall reaction, finding

$$R(\hat{\mathbf{a}}, \hat{\mathbf{b}}) \propto 2\left[1 - \frac{k^2}{E^2}\sin^2\theta\right](1 - \alpha^2 a_n b_n) + \frac{k^2}{E^2}\sin^2\theta[1 - \alpha^2(\hat{\mathbf{a}} \cdot \hat{\mathbf{b}} - 2a_x b_x)]$$

where θ is the cms scattering angle; k and E are the Λ momentum and energy, respectively; $\alpha = -0.642 \pm 0.013$ is the Λ decay asymmetry parameter resulting from parity violation; $\hat{\mathbf{a}}$ ($\hat{\mathbf{b}}$) is a unit vector along the π^+ (π^-) momentum in the $\overline{\Lambda}$ (Λ) rest frame; the x-axis is taken perpendicular to the $e^+e^- \to \Lambda\overline{\Lambda}$ scattering plane; and $\hat{\mathbf{n}}$ is the direction of polarization of the J/ψ. Törnqvist's hope to use these predictions to test Bell's inequality appears groundless, however, because (i) no dichotomic observable is directly measured, the pions' momenta being continuously variable; and (ii) momentum measurements concern compatible observables, and Landau's theorem of Section 3.2.3 shows that no violation of Bell's inequality is possible anyway. These criticisms apply also to most of the other proposals discussed in this section.

In 1992 Privitera[19] considered an EPR test of quantum mechanics based on the reaction $e^+e^- \rightarrow \tau^+\tau^-$ and on the subsequent decays $\tau^+ \rightarrow \pi^+ + \bar{\nu}_\tau$ and $\tau^- \rightarrow \pi^- + \nu_\tau$. He was well aware that a good agreement of these reactions with quantum mechanics could not exclude local hidden-variable theories in general, but only, perhaps, certain classes of them. The same reaction was considered by Abel, Dittmar, and Dreiner.[20] By studying this process, they concluded that their theoretical expressions agreed with Bell's inequality. However, their result was based on an improper use of Bell's theorem, as stressed by Datta.[21] In fact the three authors calculated the probability density

$$P(\cos \theta_{\pi\pi}) = \tfrac{1}{2}\left(1 - \tfrac{1}{3}\cos \theta_{\pi\pi}\right)$$

where $\theta_{\pi\pi}$ is the angle between the momenta of the outgoing pions of reaction (46) (after decay of Λ and $\bar{\Lambda}$). They inserted this density into Bell's inequality even though $P(\cos \theta_{\pi\pi})$ does not refer to dichotomic observables.

In 1993 Srivastava and Widom[22] published the incredible proposal that the probability of an earlier decay of a neutral kaon depends on where and when the other EPR-correlated kaon hits, at a later time, an absorber. This they claimed to be a prediction of quantum theory. Actually Ancochea and Bramon[23] showed that Srivastava and Widom had introduced a wrong sign in the *CP*-violating wave function. With the correct sign quantum theory was shown to imply no action of the future on the past.

In 1995 Di Domenico[24] published a detailed proposal for the study of the EPR paradox with a ϕ factory, based on a kaon "quasispin" formalism taking account of *CP* violation. He introduced three dichotomic ($=\uparrow, \downarrow$) variables A, B, C, treated them from the point of view of local realism according to Wigner's approach (Section 1.10), and introduced eight basic probabilities, such as

$$W_1 = W[A \uparrow (t_1), B \downarrow (t_2), C \downarrow (t_3); A \downarrow (t_1), B \uparrow (t_2), C \uparrow (t_3)]$$

This is the probability that the hidden variable assumes a value lying in the domain indicated within brackets; when this happens a measurement of the quasispin $A(t_1)$, $B(t_2)$, or $C(t_3)$ on the kaon on the left will yield a result indicated by an arrow to the left of the semicolon; a measurement on the kaon on the right will yield a result indicated to the right of the semicolon. After this correct start, "in order to simplify the notation," the author considered in W equal times for the three observables and propagated the probability in time by taking account only of decay factors. There is clearly a loss of generality in this approach, because the probabilities of the previous type are expected to vary with strangeness oscillations too. The time dependence of probabilities for correlated kaon pairs will be discussed in Section 4.4. For this reason the validity of Di Domenico's claim that in the case of kaon pairs CP violation leads to a violation of his form of Bell's inequality seems doubtful.

Another interesting proposal for the study of the EPR paradox with neutral kaon pairs has been made by Eberhard,[25] who considered an asymmetric ϕ factory in which a 2.13-GeV electron beam collides with a 122-MeV positron beam. These energies are chosen so as to correspond to a total energy in the center of mass equal to the mass of the ϕ meson, 1020 MeV. Neutral kaon pairs are produced in ϕ decays and move in the laboratory with velocities forming acute angles with the electron beam (Fig. 4.11). The test suggested is based on an interference effect between $K_S \rightarrow K_L$ regeneration processes involving two kaons, centimeters apart. Such interference seems to exhibit nonlocal features, and Eberhard shows that it leads to violations of Bell-type inequalities. Additional assumptions are, however, needed to deduce these inequalities, and this seems to weaken the whole argument, even though the author shows that his extra assumptions are rather reasonable. Anyway, no project to build asymmetric ϕ factories is known at present.

4.3.4. Direct Applications of Local Realism

The main difficulties in examining the EPR paradox experimentally with kaon pairs in usual accelerator physics are

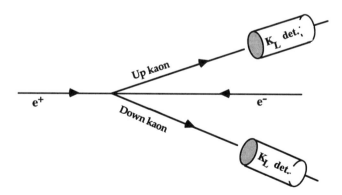

FIGURE 4.11. Neutral kaon trajectories from ϕ meson decay at an asymmetrical ϕ factory (after Eberhard[25]). The two detectors are supposed to reveal long kaons only. This constructive interference test must be carried out with regenerators set on (1) up kaon path; (2) down kaon path; (3) both paths; (4) no path.

(i) The cross sections for producing ϕ mesons are normally rather small.

(ii) Considerable background noise usually hides the signal composed of the two neutral kaons arising from ϕ decay.

Cocolicchio[26] proposed that a ϕ factory be used to overcome these difficulties. He pointed out that a machine luminosity of 10^{32} cm^{-2} s^{-1}, as expected for the DAϕNE accelerator, could easily allow one to detect the 12% discrepancy predicted in 1983 by Selleri.[9] The latter paper contained the first systematic application of local realism to kaon pairs. Elements of reality connected with strangeness and *CP* were shown to be necessarily present in each kaon. These elements of reality are the basic ingredients of any theory developed according to local realism. The quantum-mechanical prediction (40) was shown to disagree with the consequences of local realism by as much as 12%, but the simple model theory developed involved extra assumptions that undermined its generality. The four assumptions made in Ref. 9 (listed in order of decreasing reliability) were the following:

A1. *Space isotropy*: All probabilities have left–right symmetry: for all practical purposes a K_S traveling to the left and a K_L traveling to the right are physically equivalent to a K_S traveling to the right and a K_L traveling to the left.

A2. *Random decays*: Kaon decays are individual random processes at constant rates, and kaonic populations reduce exponentially.

A3. K^0–$\overline{K}^0$ *symmetry*: All probabilities have particle–antiparticle symmetry. In the physical situation described by the quantum-mechanical vector (21) one has

$$|\langle K_a^0 \overline{K}_b^0 | \psi \rangle|^2 = |\langle \overline{K}_a^0 K_b^0 | \psi \rangle|^2$$

Similar relations are assumed to hold for the local realistic probabilities.

A4. *Opposite transition rates*: It was assumed that the rate of every transition equals that of the opposite transition. Consider, for example, the transitions

$$\overline{K}_S(t)K_L(t) \rightarrow K_S(t)\overline{K}_L(t), \qquad K_S(t)\overline{K}_L(t) \rightarrow \overline{K}_S(t)K_L(t) \qquad (47)$$

where the first (second) kaon written is the one traveling to the left (right). Clearly, the two transitions (47) describe jumps in strangeness that are simultaneous in the proper times and take place for $t_a = t_b = t$. The second transition is opposite to the first.

Notice the different nature of these assumptions: space isotropy (A1) is a very natural statement supported by a large body of experimental evidence, but randomness of decays (A2), K^0–$\overline{K}^0$ symmetry (A3), and equality of the opposite transition rates (A4) are simplifying assumptions that allowed that early paper to reach a conclusion.

A discussion of the theoretical studies of the EPR paradox for $K^0\overline{K}^0$ pairs was made by Ghirardi, Grassi, and Weber (GGW).[27] Their general conclusion was

> the φ-factory facility does not seem to open new ways of testing quantum mechanics versus alternative general schemes of the type which are usually regarded as worth considering in the debate about locality and quantum mechanics.

However, the rest of the story shows that just the opposite is true, because the years following 1992 have produced new results that are decisive for a possible solution of the EPR paradox. It has been shown[11] that a φ-factory facility is the ideal tool for performing very meaningful experiments, which can decide between local realism and quantum mechanics.

It is nevertheless of some interest to review the arguments put forward by GGW. The authors started by observing that Bell's inequality written in terms of four different times of flight is not violated by the quantum-mechanical two-time joint probability for correlated pairs of kaons. This conclusion is correct and has also been reached by Cobianco,[28] who pointed out that, *a priori*, a violation is expected for large time differences, just as in the EPR experiments with atomic photon pairs it is obtained for relatively large angles between the polarizers' axes. But the decay factors in the kaonic wave functions decrease exponentially, and large times make it impossible to overcome Bell's limit. When the degree of violation is plotted as a function of the artificially varied kaon lifetime (e.g., by keeping the short/long lifetime ratio constant), one sees that for real kaons no violation exists, but that a discrepancy appears when the kaon lifetime becomes about twice as long as it really is, or more. For stable kaons one obtains the standard 41% violation.

Of course, the correctness of the GGW conclusion concerning Bell's inequality does not mean that local realism and quantum mechanics are compatible for real kaons. In fact, Bell's inequality is only one of the many consequences of local realism, as shown in Section 3.4.5. GGW were aware of this situation, for they not only examined Bell's inequality but considered proposals formulated in different terms. To Six,[14] for instance, they objected that Furry's spontaneous factorization hypothesis is not believable as a dynamical mechanism for the evolution of quantum systems because it would lead to violations of well-known conservation laws. Concerning Selleri's paper,[9] GGW argued against the generality of the local realistic approach because of assumption A4 postulating the equality of the instantaneous rates of the two processes (47). To this GGW objected that it is too restrictive, because local realistic models could be conceived in which the two rates are different. They were right about this.

The pessimistic conclusion reached by GGW concerning the usefulness of a φ factory for the study of the EPR paradox is probably due to their manifest conviction that the Orsay "clear-cut" experiments[29] have really shown that nature is nonlocal, at least regarding photon correlation experiments. Actually the logical situation is not so simple: as seen in the third chapter, the photon correlation experiments have violated only inequalities deduced with arbitrary and untestable additional assumptions that make them much stronger than Bell's original inequality. It has been impossible to test the validity of Bell's original ("weak") inequality because of the low efficiency of photon detectors. Highly efficient detectors exist instead in elementary particle physics: For this reason the kaon-pair option seems to be the best choice for a really penetrating discrimination between local realism and current quantum theory.

The seeds for overcoming the GGW objection had actually been sown one year before (1991) by Home and Selleri (HS).[10] These authors discussed only single-kaon physics (and not the EPR problem) and showed that at least one local realistic model existed that exactly reproduced the quantum-mechanical oscillations in strangeness. In this model kaons undergo sudden jumps of strangeness; quantum probabilities are reproduced by summing over several different individual behaviors. The very existence of this model eliminated the "mystery" surrounding the quantum-mechanical properties of kaons. Moreover, HS showed that the local realistic rates for the transitions $K^0 \to \overline{K}^0$ and $\overline{K}^0 \to K^0$ are necessarily different (Fig. 4.12). So the rates for transitions (46) are not expected to be equal either. It was not the task of HS to find the most general local realistic

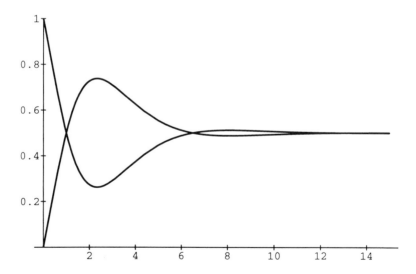

FIGURE 4.12. The results found by Home and Selleri[10] for the transition rates $K^0 \to \overline{K}^0$ and $\overline{K}^0 \to K^0$, assumed to be the same for $CP = \pm 1$. The rates are divided by Δm. Proper time in abscissa, in units γ_S^{-1}. The rate for $K^0 \to \overline{K}^0$ is the upper curve at small times.

model for single-kaon oscillations, but only to show that nothing in these oscillations contradicted local realism.

In 1992 Privitera and Selleri (PS)[30] developed the ideas introduced in Ref. 9, using a more general theoretical approach that allowed them to do without the extra assumptions A4 and A2. No effort was made, however, to avoid A3. Neglecting CP violation, PS started from the local realistic description of kaon pairs [described quantum mechanically by the state vector (21)] according to which the two particles necessarily undergo simultaneous strangeness jumps. These take place at random times in a statistical ensemble of pairs, but every jump must be predetermined for the two kaons at the time of ϕ decay to account for the simultaneity within the local realistic approach. PS developed a rate-equation method for calculating the probabilities of the eight possible local realistic states of kaon pairs. The method was so general that even the "safe" extra assumption A2 was not used. The final result was the local realistic expression of the joint probability for double $\overline{K}^0$ observations at different proper times in terms of $T_1(t)$ and $T_2(t)$. These were defined as follows:

$$T_i(t)dt = \text{probability of a transition away from the local realistic}$$
$$\text{state } \Lambda_i(t, t) \text{ in the time interval } (t, t + dt), \qquad i = 1, 2$$

where

$$\Lambda_1(t_a, t_b) = \text{local realistic state for } K_1(t_a)K_4(t_b)$$
$$\Lambda_2(t_a, t_b) = \text{local realistic state for } K_2(t_a)K_3(t_b)$$

where the notation (24) was used for the local realistic states: In the case of $\Lambda_1(t_a, t_b)$, for example, the first kaon is objectively "short" ($CP = +1$) and "particle" (strangeness $= +1$), while the second kaon is objectively "long" ($CP = -1$) and "antiparticle" (strangeness $= -1$).

The final part of the PS paper used the results for the transition rates $T_1(t)$ and $T_2(t)$ found by HS for single kaons. This approach presented an advantage and a shortcoming. The advantage was that the difference between the opposite rates was recognized, thus overcoming the dubious extra assumption A4, and hence the GGW objection. The HS model for single kaons was, however, not general enough. Extended to the theory of correlated kaon pairs this became a shortcoming and could lead only to results of limited generality. But the HS model was natural and elegant, and the discrepancy with quantum mechanics found by PS was large (up to about 25%). This result indicated that such a discrepancy was likely to remain even if the most general theory for single kaons was used. The further developments of this line of thought will be presented in the next section.

4.4. GENERAL LOCAL REALISTIC PREDICTIONS

A general local realistic theory of correlated neutral kaon pairs will now be reviewed. If kaon pairs are produced in ϕ meson decays, the local realistic probability of observing $\overline{K}^0 \overline{K}^0$ pairs at certain different proper times differs by 25–30% from the quantum-mechanical prediction, despite the fact that Bell-type inequalities are necessarily satisfied. The size of this difference justifies the systematic neglect of CP violation. A ϕ-factory accelerator seems to be the right place for a resolution of the EPR paradox, but other ways of producing ϕ mesons are not excluded.

4.4.1. The Case of Neutral Kaon Pairs

Kaon pairs arising from the decay of the ϕ meson, e.g. produced in $e^+ e^-$ collisions, are described quantum mechanically by the $J^{PC} = 1^{--}$ state vector (22). It will now be shown that the local realistic approach leads to disagreement with the probability (40) of a double $\overline{K}^0$ observation at proper times t_a and t_b. The starting point is again the discussion of Section 4.2, where it was shown that local realism applied to the physical situation described by (22) implies, at equal proper times, a total anticorrelation both in strangeness and in CP between the two kaons flying in opposite directions, the four possible physical configurations appearing initially with the same statistical weight ($\frac{1}{4}$). Given (24), one must consider the following four cases for the calculation of $P^{LR}[\overline{K}(t_a); \overline{K}(t_b)]$.

A. *Initial state with a $K_1(0)$ on the left and a $K_4(0)$ on the right.* The probability that the initial $K_1(0)$ on the left evolves into an $S = -1$ state at proper time t_a [then, given CP conservation, into $K_2(t_a)$] is given by (36):

$$p_{21}(t_a|0) = E_S(t_a)Q_-(t_a) - \rho(t_a) \tag{48}$$

Correlated with the leftward-moving antikaon $K_2(t_a)$, on the right side of the physical process there will be at time $\tilde{t}_b = t_a$ either decay products or a pure K_3 state. The probability of the latter is $E_L(\tilde{t}_b)$. The probability of its evolution into $K_4(t_b)$, conditional on the state $K_3(\tilde{t}_b)$ [with $t_b > \tilde{t}_b$], is

$$p_{43}(t_b|\tilde{t}_b) \equiv p[K_4(t_b)|K_3(\tilde{t}_b)] \tag{49}$$

Therefore, in this first case the overall probability of a double $S = -1$ observation at proper times t_a (on the left) and t_b (on the right) is clearly

$$P_1[K_2(t_a); K_4(t_b)] = p_{21}(t_a|0)E_L(\tilde{t}_b)p_{43}(t_b|\tilde{t}_b)$$
$$= \{E_S(t_a)Q_-(t_a) - \rho(t_a)\}E_L(\tilde{t}_b)p_{43}(t_b|\tilde{t}_b) \tag{50}$$

B. *Initial state with a $K_2(0)$ on the left and a $K_3(0)$ on the right.* The probability that the initial $K_2(0)$ on the left remains an $S = -1$ state at proper time t_a [then, given CP conservation, that it becomes $K_2(t_a)$] is given by (36):

$$p_{22}(t_a|0) = E_S(t_a)Q_+(t_a) + \rho(t_a) \tag{51}$$

Correlated with the leftward-moving antikaon $K_2(t_a)$, on the right side of the physical process there will be at proper time $\tilde{t}_b = t_a$ either decay products or a pure K_4 state. The probability of the latter is $E_L(\tilde{t}_b)$. The probability of its evolution into $K_4(t_b)$ is again given by Eq. (49). Therefore in this second case the overall probability of double $S = -1$ observation at proper times $\tilde{t}_b = t_a$ (on the left) and t_b (on the right) [with $t_b > \tilde{t}_b$] is

$$
\begin{aligned}
P_2[K_2(t_a); K_4(t_b)] &= p_{22}(t_a|0)E_L(\tilde{t}_b)p_{43}(t_b|\tilde{t}_b) \\
&= \{E_S(t_a)Q_+(t_a) + \rho(t_a)\}E_L(\tilde{t}_b)p_{43}(t_b|\tilde{t}_b)
\end{aligned}
\tag{52}
$$

Notice that the term $\rho(t_a)$ is going to disappear when (50) and (52) are added.

C. *Initial state with a $K_3(0)$ on the left and a $K_2(0)$ on the right.* The probability that the initial $K_3(0)$ on the left evolves into an $S = -1$ state at proper time t_a [then, given CP conservation, into $K_4(t_a)$] is given by (36):

$$p_{43}(t_a|0) = E_L(t_a)Q_-(t_a) + \rho(t_a) \tag{53}$$

Correlated with the leftward-moving antikaon $K_4(t_a)$, on the right side of the physical process there will be at time $\tilde{t}_b = t_a$ either decay products or a pure K_1 state. The probability of the latter is $E_s(\tilde{t}_b)$. The probability of its evolution into $K_2(t_b)$, conditional on its having been a $K_1(\tilde{t}_b)$ [with $t_b > \tilde{t}_b$] is

$$p_{21}(t_b|\tilde{t}_b) \equiv p[K_2(t_b)|K_1(\tilde{t}_b)] \tag{54}$$

Therefore, in this third case the overall probability of double $S = -1$ observation at proper times t_a (on the left) and t_b (on the right) is

$$
\begin{aligned}
P_3[K_4(t_a); K_2(t_b)] &= p_{43}(t_a|0)E_S(\tilde{t}_b)p_{21}(t_b|\tilde{t}_b) \\
&= \{E_L(t_a)Q_-(t_a) + \rho(t_a)\}E_S(\tilde{t}_b)p_{21}(t_b|\tilde{t}_b)
\end{aligned}
\tag{55}
$$

D. *Initial state with a $K_4(0)$ on the left and a $K_1(0)$ on the right.* The probability that the initial $K_4(0)$ on the left evolves into an $S = -1$ state at proper time t_a [then, given CP conservation, into $K_4(t_a)$] is given by (36):

$$p_{44}(t_a|0) = E_L(t_a)Q_+(t_a) - \rho(t_a) \tag{56}$$

Correlated with the leftward-moving antikaon $K_4(t_a)$, on the right side of the physical process there will be at time $\tilde{t}_b = t_a$ either decay products or a pure K_1 state. The probability of the latter is $E_S(\tilde{t}_b)$. The probability of its evolution into $K_2(t_b)$ is given by (54). Therefore, in this fourth case the overall probability of double $S = -1$ observation at proper times t_a (on the left) and $\tilde{t}_b = t_a$ (on the right) $\tilde{t}_b = t_a$ is

$$P_4[K_4(t_a); K_2(t_b)] = p_{44}(t_a|0)E_S(\tilde{t}_b)p_{21}(t_b|\tilde{t}_b)$$
$$= \{E_L(t_a)Q_+(t_a) - \rho(t_a)\}E_S(\tilde{t}_b)p_{21}(t_b|\tilde{t}_b) \qquad (57)$$

Notice that the $\rho(t_a)$ term is once more going to disappear when (55) and (57) are added.

The four elementary states of local realism must appear initially with the same weight $(\frac{1}{4})$ in the physical situation described quantum mechanically by the state vector (22), as shown in Ref. 11. Therefore, given the results obtained above, one has

$$P_{LR}[\overline{K}(t_a); \overline{K}(t_b)] = \tfrac{1}{4}\{P_1[K_2(t_a); K_4(t_b)] + P_2[K_2(t_a); K_4(t_b)] + P_3[K_4(t_a); K_2(t_b)]$$
$$+ P_4[K_4(t_a); K_2(t_b)]\}$$
$$= \frac{E_S(t_a)E_L(\tilde{t}_b)}{4}p_{43}(t_b|\tilde{t}_b) + \frac{E_L(t_a)E_S(\tilde{t}_b)}{4}p_{21}(t_b|\tilde{t}_b)$$

Remembering that $\tilde{t}_b = t_a$ the last equation can be written

$$P_{LR}[\overline{K}(t_a); \overline{K}(t_b)] = \frac{E_S(t_a)E_L(t_a)}{4}[p_{43}(t_b|t_a) + p_{21}(t_b|t_a)] \qquad (58)$$

where t_a is now used as a time label for the rightward-moving kaon as well. The probabilities p_{43} and p_{21} in (58) are not known in general: the previous considerations would fix them [up to the additive terms $\pm\rho(t_b)$] only if one had $t_a = 0$. The presence of the conditional probabilities in (58) shows that the result is valid for $t_b > t_a$ only. All the considerations of the present section could, however, be repeated with the role of the two kaons exchanged. Therefore one can conclude that

$$P_{LR}[\overline{K}(t_a); \overline{K}(t_b)] = P_{LR}[\overline{K}(t_b); \overline{K}(t_a)] \qquad (59)$$

An analogous symmetry in t_a, t_b clearly holds for the quantum-mechanical probability given by Eq. (40).

4.4.2. Two Time Probabilities

The probabilities $p_{21}(t_b|t_a)$ and $p_{43}(t_b|t_a)$ in Eq. (58) remain to be calculated. The first is the probability that a right-moving kaon, which was with certainty a K_1 at time $\tilde{t}_b = t_a$, becomes a K_2 at time t_b. The meaning of the second is similar. The time evolution mixes opposite strangeness states without changing *CP* (which is assumed to be conserved). In the case of $CP = +1$ the interesting probabilities are

$$p_{11}(t|t_0) \equiv p[K_1(t)|K_1(t_0)], \qquad p_{21}(t|t_0) \equiv p[K_2(t)|K_1(t_0)] \qquad (60)$$

which must satisfy

$$p_{11}(t_0|t_0) = 1, \qquad p_{21}(t_0|t_0) = 0$$

These probabilities are not deducible from the results of Section 4.2.4, which refer only to initial time zero, but are calculable by means of rate equations. If one defines $T_1(t)$ to be the transition rate at proper time t from K_1 to K_2, and $T_2(t)$ the opposite transition rate from K_2 to K_1; and $T_3(t)$ the transition rate at proper time t from K_3 to K_4, and $T_4(t)$ the opposite transition rate from K_4 to K_3;

$$A(t) \equiv T_1(t) + T_2(t), \qquad B(t) \equiv T_2(t) - T_1(t) \qquad (61)$$

$$\tilde{A}(t) \equiv T_3(t) + T_4(t), \qquad \tilde{B}(t) \equiv T_4(t) - T_3(t) \qquad (62)$$

$$E(t, t_0) \equiv \exp\left\{-\int_{t_0}^{t} dt'\, A(t')\right\} \qquad (63)$$

$$\tilde{E}(t, t_0) \equiv \exp\left\{-\int_{t_0}^{t} dt'\, \tilde{A}(t')\right\} \qquad (64)$$

then the probabilities $p_{21}(t_b|t_a)$ and $p_{43}(t_b|t_a)$ can be obtained by solving the rate equations and turn out to be

$$p_{21}(t_b|t_a) = E_S^{-1}(t_a)[p_{21}(t_b|0) - p_{21}(t_a|0)E_S(t_b - t_a)E(t_b, t_a)] \qquad (65)$$

$$p_{43}(t_b|t_a) = E_L^{-1}(t_a)[p_{43}(t_b|0) - p_{43}(t_a|0)E_S(t_b - t_a)\tilde{E}(t_b, t_a)] \qquad (66)$$

It is now easy to reconstruct the interesting probability (58) and obtain

$$P_{LR}[\overline{K}(t_a); \overline{K}(t_b)] = \frac{E_L(t_a)}{4}[p_{21}(t_b|0) - p_{21}(t_a|0)E_S(t_b - t_a)E(t_b, t_a)]$$
$$+ \frac{E_S(t_a)}{4}[p_{43}(t_b|0) - p_{43}(t_a|0)E_L(t_b - t_a)\tilde{E}(t_b, t_a)] \qquad (67)$$

This result is of fundamental importance because it allows one to use the single-kaon theory of Section 4.2, which described kaonic evolution starting from proper time zero. Notice that for $t_b = t_a$ one has

$$E_S(0) = E_L(0) = E(t_a, t_a) = \tilde{E}(t_a, t_a) = 1$$

From Eq. (67) it then follows that

$$P_{LR}[\overline{K}(t_a); \overline{K}(t_a)] = 0$$

The meaning of the last equation is clear: the class of local realistic theories considered satisfies the strangeness anticorrelation at equal proper times that also hold in quantum mechanics. For this reason it would not be correct to compare Eq. (67) with the "Furry" expression (43), which is in agreement with local realism but does not vanish for $t_b = t_a$.

4.4.3. Unicity of the Local Realistic Model for Single Kaons

When single-kaon physics is considered, as in Section 4.2, an unknown quantity $\rho(t)$ appears in all transition probabilities: see Eq. (36). By considering only single kaons little can be said about $\rho(t)$; only upper and lower limits can be obtained for $\rho(t)$, as in Eqs. (38) and (39). An interesting result has recently been obtained by Privitera,[31] who showed that the application of single-kaon theory to correlated kaon pairs leads necessarily to the condition

$$\rho(t) = 0 \qquad (68)$$

so that the acceptable local realistic theory of single neutral kaons becomes unique! The proof of (68) follows.

Consider again case A of Section 4.4.1 of an initial state with a $K_1(0)$ on the left and a $K_4(0)$ on the right. The probability that the initial $K_4(0)$ on the right evolves into a $S = +1$ state at proper time $\tilde{t}_b = t_a$ [which is the necessary condition for having an $S = -1$ $K_2(t_a)$ on the left] is given by (36):

$$p_{34}(\tilde{t}_b|0) = E_L(\tilde{t}_b)Q_-(\tilde{t}_b) + \rho(\tilde{t}_b)$$

Correlated with the rightward-moving kaon $K_3(\tilde{t}_b)$, on the left side of the physical process there will be at time $\tilde{t}_b = t_a$ either decay products or a pure K_2 state. The probability of the latter is $E_S(t_a)$. The probability of the further evolution of the $K_3(\tilde{t}_b)$ into $K_4(t_b)$ [with $t_b > \tilde{t}_b$] is

$$p_{43}(t_b|\tilde{t}_b) \equiv p[K_4(t_b)|K_3(\tilde{t}_b)]$$

Therefore, in case A the overall probability of a double $S = -1$ observation at proper times t_a (on the left) and t_b (on the right) can also be written

$$P_1[K_2(t_a); K_4(t_b)] = p_{34}(\tilde{t}_b|0)E_S(t_a)p_{43}(t_b|\tilde{t}_b)$$

Comparing this result with Eq. (50) one gets

$$p_{21}(t_a|0)E_L(\tilde{t}_b)p_{43}(t_b|\tilde{t}_b) = p_{34}(\tilde{t}_b|0)E_S(t_a)p_{43}(t_b|\tilde{t}_b)$$

whence, eliminating the common factor and using Eq. (36) one more,

$$[E_S(t_a)Q_-(t_a) - \rho(t_a)]E_L(\tilde{t}_b) = [E_L(\tilde{t}_b)Q_-(\tilde{t}_b) + \rho(\tilde{t}_b)]E_S(t_a)$$

Now it is enough to recall that $\tilde{t}_b = t_a$ to see that the terms containing $Q_-(t_a)$ cancel and the condition remains:

$$\rho(t_a)[E_L(t_a) + E_S(t_a)] = 0$$

Given the arbitrariness of t_a, from the last equation the result (68) follows for all t.

4.4.4. Incompatibility with Quantum Mechanics

Unknown quantities in (67) are $E(t_b, t_a)$ and $\tilde{E}(t_b, t_a)$. They can be used to deduce upper and lower bounds for the left side of (67). Since $A(t)$ and $\tilde{A}(t)$ are never negative, by definition, one gets, from (63) and (64),

$$0 \leq E(t_a, t_b), \tilde{E}(t_a, t_b) \leq 1 \tag{69}$$

In calculating the extreme values of (67) the point of view will be adopted that these functions are independent of one another. If the quantum-mechanical predictions fall outside the so-calculated set of values, a clear incompatibility between local realism and quantum mechanics will emerge.

Observing that in (67) $E(t_a, t_b)$ and $\tilde{E}(t_a, t_b)$ multiply only negative terms, one can set them equal to the extreme values 0 and 1 [see Eq. (69)] to obtain the inequalities

$$P^{\max} \geq P_{LR}[\overline{K}(t_a); \overline{K}(t_b)] \geq P^{\min} \tag{70}$$

where

$$P^{\max} = \frac{E_L(t_a)}{4} p_{21}(t_b|0) + \frac{E_S(t_a)}{4} p_{43}(t_b|0) \tag{71}$$

and

$$P^{\min} = \frac{E_L(t_a)}{4}[p_{21}(t_b|0) - p_{21}(t_a|0)E_S(t_b - t_a)]$$
$$+ \frac{E_S(t_a)}{4}[p_{43}(t_b|0) - p_{43}(t_a|0)E_L(t_b - t_a)] \tag{72}$$

By using (36) and (68) in (71), one obtains

$$P^{\max} = \frac{E_L(t_a)E_S(t_b) + E_S(t_a)E_L(t_b)}{4} Q_-(t_b) \tag{73}$$

Again using (36) and (68), written for t_b and t_a, one obtains from (72):

$$P^{\min} = \frac{E_L(t_a)E_S(t_b) + E_S(t_a)E_L(t_b)}{4}[Q_-(t_b) - Q_-(t_a)] \tag{74}$$

A calculation of the minimum predicted by local realism in Eq. (70) was performed in Ref. 11 for $t_b = 2t_a$. Those numerical results are shown in the third column of Table 4.7 and compared with $P_{QM}[\overline{K}(t_a); \overline{K}(t_b)]$: the latter probability violates the quantum-mechanical limit by as much as 30%. Privitera's result (68) was not available at that time, so $\rho(t_a)$ was also used as a variable parameter in that calculation. A better limit can now be obtained by using Eq. (74), which incorporates Eq. (68). The result is shown in the last column of Table 4.7. The values of $P^{\min}$ are thus seen to be slightly but systematically higher than

TABLE 4.7. Comparison between Predicted Probabilities for Double $\bar{K}^0$
Observations at Proper Times t_a and $2t_a$

$\gamma_S t_a$	$P_{QM}[\bar{K}(t_a), \bar{K}(2t_a)]$	$P_{LR}^{\min}[\bar{K}(t_a), \bar{K}(2t_a)]^{11}$	$P_{LR}^{\min}[\bar{K}(t_a), \bar{K}(2t_a)]$
0.2	0.0018	0.0044	0.0052
0.4	0.0051	0.0118	0.0145
0.6	0.0087	0.0171	0.0221
0.8	0.0115	0.0195	0.0258
1.0	0.0133	0.0192	0.0259
1.2	0.0142	0.0174	0.0234
1.4	0.0144	0.0148	0.0198
1.6	0.0139	0.0119	0.0158
1.8	0.0131	0.0083	0.0121

those found in Ref. 11, consistently with the larger number of theories there considered.

The incompatibility between local realism and quantum mechanics for $t_b = 2t_a$ is also shown in Fig. 4.13, where the disagreement of the quantum-theoretical predictions with the minimum probability allowed by local realism is evidenced by the values *below* unity of the quantum mechanical curve.

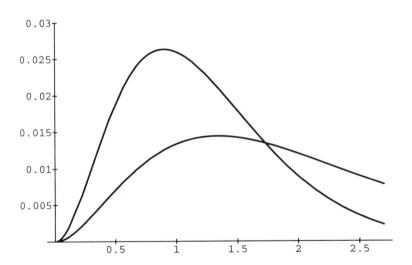

FIGURE 4.13. Predictions of quantum mechanics and local realism [$P^{\min}$, given by Eq. (74)] for the probability of a double $\bar{K}^0$ detection, for $t_b = 2t_a$. Time t_a in abscissa (units γ_S^{-1}). The quantum-mechanical curve is lower than the local realistic minimum for $t_a < 1.7$.

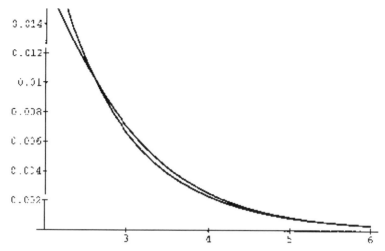

FIGURE 4.14. Comparison between the predictions of quantum mechanics and local realism [maximum, given by Eq. (73)] for the probability of a double $\bar{K}^0$ detection ($t_b = 2.5t_a$). Time t_a in abscissa (units γ_S^{-1}). For $t_a > 2.6$ the quantum-mechanical curve becomes larger than the local realistic maximum.

A similar disagreement exists with the maximum local realistic probability: see Fig. 4.14, relative to the case $t_b = 2.5t_a$, where the quantum-theoretical violations of local realism are evidenced by the values of the quantum mechanical curve being above those of the maximal local realistic curve.

In the plane of the proper times t_a and t_b of the two kaons the regions of quantum-mechanical violations of the maximum and minimum deduced from local realism for the double $\overline{K}^0$ detection probability are the regions R_1 and R_2 in Fig. 4.15. Only the part with $t_a < t_b$ is shown in the figure: since the theory is completely symmetrical for $t_a \leftrightarrow t_b$, one can easily imagine a similar situation in the region above the line $t_a = t_b$ of the (t_a, t_b) plane.

During the preparation of R. Foadì's Thesis for the master degree at the University of Turin (1998) it became clear that the treatment of kaon pairs based on rate equations as given in Ref. 11 can lead to inconsistencies when applied to a broader extent. In order to overcome this difficulty one needs to take into careful account the historical nature of kaons: in a way, they remember their past. When this is done properly an even larger discrepancy (up to 73%) is found between local realism and quantum mechanics. This forms the object of a paper in preparation.[32]

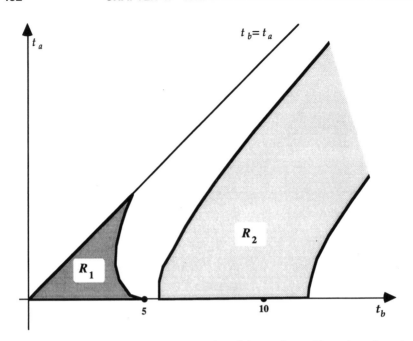

FIGURE 4.15. Plane of the proper times t_a and t_b of the two kaons. The regions of quantum-mechanical violations of the maximum and minimum deduced from local realism for the double $\bar{K}^0$ detection probability are roughly regions R_1 and R_2. Violations of the minimum (maximum) are inside R_1 (R_2). Only the part with $t_a < t_b$ is shown, the theory being symmetrical for $t_a \leftrightarrow t_b$. Time t_b in abscissa (units γ_S^{-1}).

REFERENCES

1. Particle Data Group, *Phys. Lett.* B **239**, 1–516 (1990).
2. C. N. YANG, *Phys. Rev.* **77**, 242–245 (1950).
3. L. R. KASDAY, in: *Proceedings of the International School of Physics "Enrico Fermi,"* Course IL (B. D'ESPAGNAT, ed.), Academic Press, New York (1971), pp. 195–210; L. R. KASDAY, J. D. ULLMAN, and C. S. WU, *Nuovo Cim. B* **25**, 633–661 (1975).
4. G. FARACI, D. GUTKOWSKI, S. NOTARRIGO, and A. S. PENNISI, *Nuovo Cim. Lett.* **9**, 607–611 (1974).
5. A. R. WILSON, J. LOWE, and D. K. BUTT, *J. Phys. G* **2**, 613–624 (1976).
6. M. BRUNO, M. D'AGOSTINO, and C. MARONI, *Nuovo Cim. B* **40**, 143–152 (1977).
7. G. BERTOLINI, E. DIANA, and A. SCOTTI, *Nuovo Cim. B* **63**, 651–665 (1981).
8. M. LAMEHI-RACHTI and W. MITTIG, *Phys. Rev. D* **14**, 2543–2555 (1976).
9. F. SELLERI, *Nuovo Cim. Lett.* **36**, 521 (1983).
10. D. HOME and F. SELLERI, *J. Phys. A* **24**, L1073–L1078 (1991).
11. F. SELLERI, *Phys. Rev. A* **56**, 3493–3506 (1997).
12. H. J. LIPKIN, *Phys. Rev.* **176**, 1715–1718 (1968).

13. R. BALDINI CELIO, M. E. BIAGINI, S. BIANCO, F. BOSSI, M. SPINETTI, A. ZALLO, and S. DUBNICKA, in: *Proc. Workshop on Physics and Detectors for DAφNE* (G. PANCHERI, ed.), INFN, Frascati (1991), pp. 179–188.

14. J. SIX, *Phys. Lett. B* **114**, 200–202 (1982).

15. O. PICCIONI in: *The Physics Handbook* DAφNE (L. MAIANI *et al.*, eds.), INFN, Frascati (1992), pp. 279–290.

16. A. DATTA and D. HOME, *Phys. Lett. A* **119**, 3–6 (1986).

17. A. MULLER, private communication; See also: CPLEAR Collaboration (A. Apostolakis *et al.*), *Phys. Lett. B* **422**, pp. 339–348 (1998).

18. N. A. TÖRNQVIST, *Phys. Lett. A* **117**, 1–4 (1986); in: *Quantum Mechanics versus Local Realism. The Einstein–Podolsky–Rosen Paradox* (F. SELLERI, ed.), Plenum, New York (1988), pp. 115–132.

19. P. PRIVITERA, *Phys. Lett. B* **275**, 172–180 (1992).

20. S. A. ABEL, M. DITTMAR, and H. DREINER, *Phys. Lett. B* **280**, 304–312 (1992).

21. A. DATTA, Proceedings of the Workshop on High Energy Physics Phenomenology, Calcutta (1997).

22. Y. SRIVASTAVA and A. WIDOM, *Phys. Lett. B* **314**, 315–319 (1993).

23. B. ANCOCHEA and A. BRAMON, *Phys. Lett. B* **347**, 419–423 (1995).

24. A. DI DOMENICO, *Nucl. Phys. B* **450**, 293–324 (1995); Testing Quantum Mechanics at DAφNE, paper presented at the Workshop on *K* physics, Orsay, France, May 30–June 4, 1996.

25. P. H. EBERHARD, *Nucl. Phys. B* **398**, 155–183 (1993); Tests of Non-local Interferences in Kaon Physics at Asymmetric φ-factories, LBL-33937 (1993).

26. D. COCOLICCHIO, *Found. Phys. Lett.* **3**, 359–365 (1990).

27. G. C. GHIRARDI, R. GRASSI, and T. WEBER, in: *The DAφNE Physics Handbook*, vol. I (L. Maiani *et al.*, ed.), INFN, Frascati (1992), pp. 261–278.

28. S. COBIANCO, Thesis, University of Bologna (1996).

29. A. ASPECT, P. GRANGIER, and G. ROGER, *Phys. Rev. Lett.* **47**, 460–463 (1981); **49**, 91–94 (1982); A. ASPECT, P. DALIBARD, and G. ROGER, *Phys. Rev. Lett.* **49**, 1804–1807 (1982).

30. P. PRIVITERA and F. SELLERI, *Phys. Lett. B* **296**, 261–272 (1992).

31. P. PRIVITERA, private communication.

32. R. FOADI and F. SELLERI, paper in preparation (1998).

Chapter 5

Proposed Solutions of the Paradox

5.1. ACTION AT A DISTANCE

There have been various attempts to *explain* the long-range interference effects predicted by quantum theory which appear to have been confirmed by experiments. Such attempts often involve a subquantal medium or ether that transmits the influence superluminally.

5.1.1. Wholeness and Holograms

According to Bohm, the essential new feature implied by quantum theory is nonlocality: a system cannot be broken up into parts whose basic properties do not depend on the state of the whole system, no matter how far apart they may be. He believes, with Hiley, that the experiments to test Bell's inequality clearly reveal the nonlocal nature of quantum phenomena.[1] Indeed nonlocality

> ... is involved in an essential way in every manifestation of a many-body system, as treated by Schrödinger's equation in a 3N-dimensional configuration space.[1]

It is the quantum potential that gives rise to the nonlocal nature of quantum phenomena. Bohm[2] has also suggested an ontological model for quantum nonlocality to make the influences in question more plausible. He has done this by introducing the new notion of "unbroken wholeness" that characterizes two correlated quantum systems. He considers the interesting example of a hologram and stresses that the different parts of the object are not in correspondence with different parts of its hologram, but rather each of the latter parts individually somehow expresses the whole object. Accordingly, by illuminating any part of the hologram, information can be obtained about the whole object, even if less detailed and from fewer angles. What can appear to be two separated

195

quantum objects may likewise be a manifestation of an interconnected wholeness. The hologram of two spheres, for instance, stores the information of each ball over the entire hologram. It can therefore be said that *in the hologram* the two spheres are really in a way amalgamated and inseparable. Bohm views this as an example of the true physical situation giving rise to the EPR paradox: In space there is only an "unbroken wholeness," which sometimes can give rise to manifestations that *appear* to be two separate objects.

As a further example, Bohm considers a device consisting of two concentric glass cylinders, separated by a highly viscous fluid such as glycerine, which is so designed that the outer cylinder can be turned very slowly, thus causing the diffusion of the viscous fluid to be negligible. If one puts a droplet of insoluble ink into this viscous fluid and slowly turns the outer cylinder, the ink will become strung out into a thread that finally becomes invisible. When the rotation of the cylinder is reversed, the ink will suddenly become visible again. Now, it could be said that the ink particles have been *enfolded* into the glycerine, like an egg is enfolded into a cake. But while one can in no way unfold the egg, the material of which undergoes irreversible diffusive mixing, it is possible to *unfold* the ink droplet from the glycerine, because in this case the mixing is viscous, diffusion being prevented by the slow turning of the outer cylinder.

Now suppose that, instead of a single ink droplet, one starts with two spatially separated droplets. When the outer cylinder is turned, each droplet will, as before, become enfolded until finally particles from both droplets appear to intermingle at random. Despite appearances, however, the ink particles retain their identification by origin because they are correspondingly drawn out into two threads that remain separate and distinct, indeed in such a way that, on reversal of the fluid motions, the threads will unfold ultimately to reconstitute the respective droplets from which they originated. With the help of this analogy, Bohm draws a distinction between what he calls the *enfolded order* and the undisguised, or unfolded, order, which constitutes our ordinary description of reality.

In a situation where the threading out of "droplets," as before, exists, one must make a distinction between wholes according to where the process of unfolding would produce a droplet or whether it would produce two droplets, and so on (different states of the vacuum). It is possible to have situations where what is going to be visible is only a very small part of the enfolded order, and one therefore introduces a distinction between what is manifest and what is not. Reality enfolds and becomes nonmanifest, or it might unfold to reveal a manifest order, and then enfold again. Bohm claims that, whereas the fundamental movement of Descartes is one that crosses space in time, a localized entity moving from one place to another, his own fundamental movement is marked by folding and unfolding. In this theory it is obviously possible to have interactions between two entities that manifest themselves as two separated objects, and in principle this view would resolve the EPR paradox.

5.1.2. Superluminal Motions in an Ether of Rigid Particles

A second possible solution of the EPR paradox is the nonlocal model of Vigier and collaborators.[3] They adopt the idea, first proposed by Dirac,[4] that the ether, with suitable properties, is not ruled out by special relativity, especially if account is taken of the probabilistic nature of quantum phenomena. It is assumed that the velocity distribution of the particles constituting the ether has a constant value over the hyperboloid

$$v_0^2 - v_1^2 - v_2^2 - v_3^2 = 1$$

In such a case, in fact, the velocity distribution looks the same to all observers, and the ether does not produce any physical effect on moving bodies. In Vigier's model this etherlike physical vacuum is made of extended *rigid* particles that can support within their interiors signals with superluminal velocity. The statistical properties of quantum objects then merely reflect the real random fluctuations of the ether.

In this theory there are also quantum waves that propagate as real physical collective excitations (i.e., as density waves) on top of Dirac's ether. In this way information originating on the boundary of the ψ-wave (such as the opening or closing of a slit in the double-slit experiment, or the observation of one of the two particles forming an EPR pair) reacts with superluminal velocity (via the quantum potential) on the particle motions that propagate with subluminal group velocities along the flow lines of the quantum-mechanical ψ-waves.

In Vigier's opinion the existence of superluminal propagation does not necessarily imply a breakdown of causality, if "causality" is defined as follows:

1. Possibility of solving the two-particle problem in the forward (or backward) time direction as a Cauchy problem;
2. Timelike nature of all particle trajectories;
3. Invariance of the formalism under the Poincaré group of transformations.

The following consequence can be drawn from a theory with superluminal connections. A particle on Earth and another in Andromeda form an EPR pair whose deterministic evolution is governed by equations containing nonlocal potentials such as Bohm's. Either one finds out what the other is doing right away and reacts accordingly. If the particle on Earth is subjected to a magnetic field, say, the other will respond instantaneously via the superluminal physical connection. The particle in Andromeda could, for instance, hit a detector if (and only if) the magnetic field in question is on. Using ensembles of correlated EPR pairs, it then becomes possible to transmit instantaneous information from Earth to Andromeda.

5.1.3. Nonlocal Sea of Microlevel Potentialities

Stapp believes that the quantum-mechanical predictions for the situation dealt with in the EPR paradox have been accurately confirmed "under experimental conditions essentially equivalent to those needed for the EPR argument." Hence he concludes that the world we live in is nonlocal. However, he does not believe that the results obtained by Bell[5] and by Clauser and Horne[6] establish that nonlocality is required because, in his opinion, these authors made very strong assumptions about microscopic reality that are not compatible with orthodox quantum thinking. The refusal of these "strong" assumptions of realism does not imply, however, any retreat to idealism or subjectivism. They can be replaced by a variant of the Copenhagen interpretation.

Stapp distinguishes between a strict Copenhagen interpretation, in which nothing is said about any reality lying beyond our observations, and an informal interpretation, in which one accepts the commonsense idea of a macroscopic reality that exists independently of our observations and can be described, at least approximately, with the concepts of classical physics. This "informal" interpretation is at least partly related to Heisenberg's idea of a transition from the "possible" to the "actual" taking place during the act of measurement. Stapp's microworld is a "sea of micro-level potentialities" that become "well-defined" physical properties only by interacting with an experimental apparatus.

A model theory proposed by Stapp[7] contains certain "hidden variables" λ, which represent all the deterministic and stochastic quantities that characterize the unified organic world and that are not used to provide the basis for a factorization structure of probabilities; they do not reflect ideas of separation, localization, or microscopic structure. Stapp writes $\lambda = (\lambda', \lambda'')$, where λ' is strictly predetermined and λ'' is any stochastic variable.

It is also assumed that every act of measurement involves a choice, which picks the actual from among what had previously been mere possibilities: The choice renders fixed and settled something that had prior to the choice been undetermined. A "choice" variable Z is also introduced and written $Z(x, y)$, where x and y represent the choices of experiment in the regions R_α and R_β, respectively, where two correlated observations of EPR type are made. The "choices" x and y are treated as independent free variables. Each can assume an infinite number of different values.

Suppose the observables A and A' can be measured on the α-particle, and B and B' on the β-particle. The choice variable picks one observable before measurement. More precisely, the chosen observable in R_α is

$$A \quad \text{if } x \in X; \qquad A' \quad \text{if } x \in X'$$

where $X \cup X'$ is the set of possible values of x. Symmetrically, in R_β the chosen observable is

$$B \quad \text{if } y \in Y; \qquad B' \quad \text{if } y \in Y'$$

where $Y \cup Y'$ is the set of possible values of y. So four possible joint measurements can be performed, depending on the values assumed by x and y: (A, B); (A, B'); (A', B); (A', B'). The outcomes of whatever measurements have been chosen are assumed to be $r_\alpha(x, y, \lambda)$ in R_α and $r_\beta(x, y, \lambda)$ in R_β. A dependence of r_α on y and of r_β on x is clearly nonlocal; in a local theory r_α, for instance, cannot depend on a choice made arbitrarily far away in R_β.

Stapp could easily prove that the local choice contradicts the empirical predictions of quantum theory and concluded:

> Neither determinism, nor counterfactual definiteness, nor any idea of reality incompatible with orthodox quantum thinking need be assumed in order to prove the incompatibility of the empirical predictions of quantum theory with the EPR idea that no influence can propagate faster than light.

The remark about the absence of a "counterfactual definiteness" is justified by the important fact that in Stapp's theory the choice of x and y and λ fixes the value *only of the observable that is actually measured*; the values of the other three observables remain completely indefinite.

5.2. RETROACTIONS AND ERGODICITY

5.2.1. The Victory of Formalism over Modelism

The possibility of modifying the past by means of retroactions from the future was first proposed as a solution to the EPR paradox by Costa de Beauregard.[8] He noted that twice in classical physics contradictions between factlike irreversible processes and the lawlike reversibility of the physical theory had been discovered.

(i) When Boltzmann used statistical mechanics for deducing the Second Law of Thermodynamics: The paradox inherent in extracting time asymmetry from a theory like Newtonian mechanics that is intrinsically time symmetric was exposed in specific forms by Loschmidt and Zermelo.

(ii) When the principle of retarded waves was used in physical optics and in classical electrodynamics in order to exclude one-half of the mathematically permissible solutions of the wave equations.

Costa de Beauregard's idea is that a careful examination of nature is bound to lead to the conclusion that retroactions in time do play a role and should not be discarded as in (i) and (ii). One way to see this is to remember that for Aristotle, to whom the concept is due, information was not only knowledge, but organizing power. The examples he gave were the craftsman's or artist's work and biological ontogenesis. A second way to see a final cause at work is to consider modern cybernetics, which, surprisingly, rediscovered the two faces of Aristotle's "information." In computers and other information-processing machines the chain

$$\text{information} \overset{(1)}{\to} \text{negentropy} \overset{(2)}{\to} \text{information}$$

means that a concept is coded and sent as a message, before being decoded and received. Negentropy is negative entropy. Step (2) is the *learning transition*, where information represents an increase in knowledge, while step (1) is the *willing transition*, in which information appears as organizing power.

In the theoretical framework (*de jure*) there is a complete symmetry between the two transitions. In spite of this there is a disymmetry in practice (*de facto*) because irreversibility is generated by misprints in the coding, noise along the line, mistakes in decoding, and so on.

The relationship between the variation of negentropy ΔN and the variation information ΔI is

$$\Delta N = k \ln 2 \Delta I$$

If N and I are both expressed in "practical" units, it turns out that the factor multiplying ΔI is very small, about 10^{-16}. Therefore, concludes Costa de Beauregard, it is very difficult to produce important increases of negentropy (decreases of entropy) by increasing the information. *Vice versa*, even a very small increase of negentropy can give rise to a large gain of information. If one lets $k \to 0$, one obtains a situation where gaining knowledge is absolutely costless, but producing order is impossible. In this limit consciousness is made totally passive; it registers what is going on without it, and does no more.

If the roots of Costa de Beauregard's conceptions go deep into classical physics, it is in quantum theory that he thinks the most important effects of retroaction can be seen. Again, he stresses, the theory is completely time symmetrical, but only until measurement causes the *collapse of the wave function*. At this point quantum theory commits itself to the philosophy of retarded waves. In Costa de Beauregard's opinion this happens because "the Copenhagen school has forgotten the hidden face of Aristotle's information."

It is precisely in the situations envisaged by the EPR paradox that his "hidden face" shows up again. In order to understand the essence of the EPR paradox, Costa de Beauregard considers the mathematical apparatus of quantum theory and concludes that the problem is today only that of tailoring the wording of the EPR situation to suit the mathematics; in his opinion there has, in the twentieth century, been an irreversible victory of formalism over modelism.

From this starting point he deduces that when an EPR pair, for instance two photons described by a nonfactorizable state vector, is measured by two observers in two regions separated by a spacelike distance, it is precisely the act of observation that produces the right physical properties of the photon pair in the past of the measurement process. Each observer is thus considered capable of telediction plus teleaction, by taking, as it were, a relay in the past, more precisely in the source that emitted the two photons.

According to this theory the element of reality introduced in the formulation of the EPR paradox can be considered real, but ought to be viewed as being created by actual acts of observation; it then propagates backward in time with one of the two correlated quantum objects from the region of measurement to the source.

In particular, there can be no question of associating elements of reality with unmeasured observables, as done originally by EPR, and later by Bell and by other authors. In this sense the solution of the EPR paradox proposed by Costa de Beauregard is similar to Bohr's. Others, such as Stapp,[9] Davidson,[10] Rayski,[11] Rietdijk,[12] Cramer,[13] and Sutherland,[14] have proposed similar solutions.

5.2.2. A Polarizable Fundamental Medium

According to the nonergodic interpretation of quantum mechanics, a sequence of quantum objects, even if separated by large time intervals, do not behave independently in their interactions with the measuring apparatus. These objects may interact with one another by means of memory effects in a hypothetical medium, filling the space they cross on their way toward the measuring instruments.

Consider, for instance, the two-slit experiment. The indirect interaction in question is such that a particle passing through a slit knows if the other slit is open, because this is somehow recorded in the medium filling the space between the two screens. Those particles that came from the second slit modified the physical properties of space and gave rise to the storage of the information in question. Obviously, interference can happen only after a sufficiently large number of particles have crossed the apparatus and conditioned the medium. In this way particles interfere with other particles, but only indirectly through the medium.[15,16]

More generally, consider a quantum experiment repeated many times, every repetition being called a "run." Let R represent the number of runs and N the number of quantum objects in every run, assumed constant for simplicity. Let λ_{rn} represent the state of the nth particle in the rth run and s_{rn} the state of the experimental apparatus just before interacting with the nth particle of the rth run. The result of the measurement, A_{rn}, is assumed to be completely fixed once λ_{rn} and s_{rn} are given. Therefore,

$$A_{rn} = A(\lambda_{rn}, s_{rn}) \tag{1}$$

Starting from these numbers two types of averages are possible:

$$\bar{A}_r = \frac{1}{N} \sum_{n=1}^{N} A_{rn}, \qquad \bar{A}_n = \frac{1}{R} \sum_{r=1}^{R} A_{rn} \tag{2}$$

Here, $\bar{A}_r$ is called the *run average*, $\bar{A}_n$ is the *ensemble average* at "time" n. Buonomano[15] observed that it is always implicitly or explicitly assumed that $\bar{A}_r = \bar{A}_n$ (*ergodic assumption*) but that such an assumption should really be checked with suitable experiments. In order to do so it is clear that the only possibility of avoiding the medium polarization effects is to keep the runs distant in time from one another, and eventually in different regions of space where no experiments have been carried out. Thus, the ensemble average for $n = 1$ is $\bar{A}_n = 1$, which should represent events collected in conditions where the medium does not act on the particles (no memory effects for $n = 1$, since no previous particles entered the apparatus in each of the considered runs!). Therefore $\bar{A}_{n=1}$ should describe a situation in which no quantum phenomenon manifests itself and classical physics holds unreservedly. Instead, $\bar{A}_n$ for large n and $\bar{A}_r$ for all r describe quantum-mechanical situations. The case of $\bar{A}_n$ for not too large values of n, but with $n \neq 1$, represents mixed situations where a transition between classical and quantum physics is taking place.

This nonergodic interpretation of quantum mechanics can, in principle, solve the EPR paradox because it can explain the apparent violations of local realism due to nonergodic effects within a strictly local theory. Consider the left side of a polarization–correlation experiment and divide the space between polarizer and source into M cells, numbering them from left to right. Thus, the polarizer is in cell 1 and the source is in cell M. Assume that the state of cell M depends on the previous state of the neighboring cells. It follows that after one photon has passed, the state of cell 2 depends on the state of the polarizer. After two photons have passed, cell 3 depends on the state of the polarizer, etc. Then after $n \geq M$ photons have passed, cell M (that is, the source) depends on the state of the polarizer.

If the right side of the polarization–correlation experiment is treated in the same manner, one obtains a situation in which the source produces pairs of photons in a state dependent on the configuration of the analyzing–detecting apparatus. As is well known, no Bell-type inequality can be obtained in such a case, and the EPR paradox does not exist.

5.2.3. Generalized Probabilities

The idea of negative probabilities has been entertained by physicists, including Dirac and Feynman. In 1942 Dirac expressed the opinion that

> Negative energies and probabilities should not be considered as nonsense. They are well-defined concepts mathematically, like a negative sum of money, since the

equations which express the important properties of energies and probabilities can still be used when they are negative. Thus negative energies and probabilities should be considered simply as things which do not appear in experimental results.[17]

In 1982 Feynman[18] stated that the only difference between a probabilistic classical world and the quantum world "is that somehow or other it appears as if the probabilities would have to go negative...."

Following these ideas, a "negative-probability solution" of the EPR paradox has been proposed by Mückenheim[19] (see also Maddox[20]). In order to understand the logical possibility of solving the EPR paradox by extending the range of variation of probabilities, one should remember that in the proofs of Bell's inequality the implicit assumption is always made that probabilities (and frequencies in ensembles) are positive and never larger than unity. For example, in Wigner's proof of Bell's inequality the probabilities $\rho(s, s'; t, t')$ were introduced [see Eq. (75) of Chapter 1], which were, by definition, positive and not larger than unity. Similarly, the proof based on factorizable probabilities used in an essential way the inequalities $0 \leq x, x', y, y' \leq 1$, where x, x', y, y' were later identified with probabilities. If these conditions are relaxed in both examples, the validity of Bell's inequality no longer follows.

In view of these considerations it is perhaps not surprising that Mückenheim could build a negative-probability *local* hidden-variable model that reproduces all the predictions of quantum theory for the "singlet" state of two spin-$\frac{1}{2}$ particles.

The two particles have a spin vector, $\mathbf{S}$ for the first one and $-\mathbf{S}$ for the second one, where $\mathbf{S}$ is assumed to have a random distribution over the sphere of radius $(\sqrt{3}/2)\hbar$ in a statistical ensemble of such pairs. The length $(\sqrt{3}/2)\hbar$ is chosen in such a way as to reproduce the quantum-mechanical eigenvalue of $\mathbf{S}^2$, which is $\frac{3}{4}\hbar^2$. If $\hat{\mathbf{a}}$ is a unit vector, the projection of $\mathbf{S}$ over $\hat{\mathbf{a}}$ satisfies

$$-\frac{\sqrt{3}}{2}\hbar \leq \mathbf{S}\hat{\mathbf{a}} \leq +\frac{\sqrt{3}}{2}\hbar \tag{3}$$

Next Mückenheim assumes that the probabilities $\omega(\hat{\mathbf{a}}+, \mathbf{S})$ and $\omega(\hat{\mathbf{a}}-, \mathbf{S})$ of measuring $\mathbf{S}\cdot\hat{\mathbf{a}}$ and finding the positive and the negative eigenvalue, respectively, are linear functions of $\mathbf{S}\cdot\hat{\mathbf{a}}$ and that their expressions satisfying

$$\omega(\hat{\mathbf{a}}+, \mathbf{S}) + \omega(\hat{\mathbf{a}}-, \mathbf{S}) = 1 \tag{4}$$

are

$$\omega(\hat{\mathbf{a}}+, \mathbf{S}) = \frac{1}{2} + \frac{\mathbf{S}\cdot\hat{\mathbf{a}}}{\hbar} \quad \text{and} \quad \omega(\hat{\mathbf{a}}-, \mathbf{S}) = \frac{1}{2} - \frac{\mathbf{S}\cdot\hat{\mathbf{a}}}{\hbar} \tag{5}$$

Obviously these probabilities can assume negative values because of (3).

For an EPR pair one can consider the case of correlated spin measurements along $\hat{\mathbf{a}}$ and $\hat{\mathbf{b}}$ for the first and second particle, respectively. The correlation function is

$$P(\hat{\mathbf{a}}, \hat{\mathbf{b}}) = \frac{\hbar^2}{16\pi} \int d\Omega \, [\omega(\hat{\mathbf{a}}+, \mathbf{S}) - \omega(\hat{\mathbf{a}}-, \mathbf{S})][\omega(\hat{\mathbf{b}}+, -\mathbf{S}) - \omega(\hat{\mathbf{b}}-, -\mathbf{S})] \quad (6)$$

Substituting (5) in (6) and integrating give

$$P(\hat{\mathbf{a}}, \hat{\mathbf{b}}) = -\frac{\hbar^2}{4} \hat{\mathbf{a}} \cdot \hat{\mathbf{b}}$$

which coincides with the quantum-mechanical correlation function for the singlet state. A local model is thus able to reproduce the quantum-mechanical violations of Bell's inequality if negative probabilities are introduced.

It has also been shown that the introduction of complex probabilities in the EPR paradox can reconcile locality with the quantum-mechanical predictions.[21] Negative probabilities have even been invoked,[22] in the context of kaon pairs involuntarily.[23]

5.3. VARIABLE PROBABILITY OF DETECTION

The idea of "variable probabilities" as a solution of the EPR paradox starts from the evidence provided by the performed experiments with atomic photon pairs and assumes that the inequalities of strong type (deduced from local realism and additional assumptions) are violated.

The point of view adopted with this line of research is that the additional assumptions, not local realism, should be blamed for the failure of the strong inequalities. One must then study local models of reality in which the logical negation of the additional assumptions is explicitly taken as true. The interesting models should thus imply the simultaneous validity of the following three statements:

1. Given that a pair of photons emerges from two regions of space where two polarizers can be located, the probability of their joint detection from two photomultipliers depends on the presence and/or orientation of the polarizers ($\overline{\text{CHSH}}$ property).
2. For a photon in state λ the probability of detection with a polarizer in place on its trajectory can be larger than the detection probability with the polarizer removed ($\overline{\text{CH}}$ property).
3. For a photon in state λ the sum of the detection probabilities in the "ordinary" and "extraordinary" beams emerging from a two-way polarizer depends on the polarizer's orientation ($\overline{\text{GR}}$ property).

From a general point of view one can maintain that local realism cannot be refuted by experiments designed for testing the strong inequalities. Only if weak inequalities could be tested could a crucial confrontation between quantum theory and local realism finally take place. This appears unlikely in the foreseeable future as far as experiments with pairs of atomic photons are concerned. This situation is, however, much better for some proposed particle physics experiments discussed in Chapter 4, since detectors are in those cases nearer to the ideal behavior.

Even in the case of low-efficiency detectors there are interesting investigations to be carried out, for example by replacing the usual additional assumptions with more physical restrictions. After all, it is unlikely that the large disagreement between quantum theory and local realism for high-efficiency detectors becomes a perfect agreement for low-efficiency detectors! For example, it would be interesting to study the use of symmetrical functions for describing the detection processes of the two photons, since it has been shown by Caser[24] that the quantum-theoretical predictions cannot in such a case agree with the factorizable probabilities.

5.3.1. A Particular Local Model

Bell's inequality[5] can be written in several ways, for instance in terms of the coincident probability $\omega(a, b)$ that the two photons of an EPR-type experiment are detected by two counters after crossing two polarizers with axes a and b. It then reads

$$-1 \le \omega(a, b) - \omega(a, b') + \omega(a', b) + \omega(a', b') - t_1(a') - t_2(b) \le 0 \qquad (7)$$

where $t_1(a')$ $[t_2(b)]$ is the probability that the first (second) photon is detected after crossing a polarizer with axis a' (with axis b), irrespective of what happens to the second (first) photon.

Now the quantum-mechanical predictions are

$$\omega(x, y) = \tfrac{1}{4}[(\epsilon_M + \epsilon_m)^2 + (\epsilon_M - \epsilon_m)^2 F \cos 2(x - y)]\eta_1 \eta_2$$

$$t_1(a') = \tfrac{1}{2}(\epsilon_M + \epsilon_m)\eta_1 \qquad (8)$$

$$t_2(b) = \tfrac{1}{2}(\epsilon_M + \epsilon_m)\eta_2$$

where ϵ_M and ϵ_m are well-known parameters related to the efficiencies of the polarizers, F is a geometrical factor, and η_1 and η_2 are the quantum efficiencies of the two counters.

Although the quantum efficiencies for the (ideal) case in which

$$\epsilon_M = 1, \quad \epsilon_m = 0, \quad F = 1, \quad \eta_1 = \eta_2 = 1$$

are known to violate Bell's inequality, such efficiencies are currently impossible. The parameters ϵ_M, ϵ_m, and F are close to their ideal values, but the quantum efficiencies of the two counters are much less than unity. Typical values are $\eta_1 \simeq \eta_2 \simeq 0.15$. It is very easy to see that inequality (7) cannot be violated by (8) if η_1 and η_2 have such values. This point was discussed at length in Chapter 3.

Here a detailed model of the photon with variable detection probability (VDP) will be constructed which: agrees exactly with the usual quantum theory for single-photon experiments; agrees very closely (within the existing experimental errors, although not exactly; see Caser[24]) with quantum theory for the EPR-type experiments so far performed; leads to new predictions if the experimental apparatus of the latter experiments is somewhat modified.

Every photon is assumed to possess a physical vector l perpendicular to its direction of motion. If the photon impinges on a polarizer with axis oriented along the direction $\hat{a}$, the *local* interaction between photon and polarizers is assumed to be such that if the angular distance of l from $\pm\hat{a}$ is less than $45°$ the photon is transmitted with certainty; otherwise it is absorbed with certainty. Figure 5.1 represents the photon transmitted with certainty.

In general if l belongs to the hatched region R_a of Fig. 5.1 it will be transmitted; if it is outside R_a it will be absorbed. A precise mechanism producing transmission or absorption for given l and $\hat{a}$ can be freely invented by the reader; only one feature of it interests us here—the fact that it is local; in other words, it depends only on the given photon, the polarizer, and their mutual physical interaction. A natural assumption is that a beam of N "unpolarized" photons is represented by an ensemble of l-vectors uniformly distributed over 2π. Since the region R_a of Fig. 5.1 covers exactly half of the possible directions, there will be, on average, $N/2$ photons transmitted and $N/2$ absorbed and the transmission probability will be $\frac{1}{2}$ in agreement with experiment.

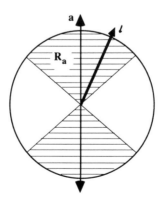

FIGURE 5.1. Hidden-variable model for photons. A photon with polarization vector l and interacting with a polarizer with axis $\hat{a}$ is transmitted if l lies in region R_a, and absorbed otherwise.

Obviously this model, like all hidden-variable models, contains *dispersion-free states*, which are those describing a statistical ensemble of (single) photons, all with the same l: for such dispersion-free ensembles the choice between transmission and absorption is fixed for all conceivable polarizer orientations.

Quantum states can also be defined, in the following way: a statistical ensemble of photons with l-vectors all contained within the region R_a of Fig. 5.1 is such that all photons will be transmitted with certainty through a polarizer with axis $\hat{a}$. Moreover this is true of the l angular density, which gives the right probability for transmission through two consecutive polarizers, the first with axis $\hat{a}$ and the second with axis $\hat{a}'$. It is given by

$$\rho_1(l' - a) = \tfrac{1}{2}\cos 2(l' - a), \qquad l' \in R_a$$
$$= 0, \qquad\qquad l' \notin R_a \tag{9}$$

Notice that $\cos 2(l' - a)$ is never negative in R_a (that is, for $-\pi/4 \le l' - a \le \pi/4$ and for $-3\pi/4 \le l' - a \le 5\pi/4$) and can therefore represent a probability. Furthermore,

$$\int_{R_a} dl' \, \frac{1}{2}\cos 2(l' - a) = 2 \int_{a-\pi/4}^{a+\pi/4} dl' \, \frac{1}{2}\cos 2(l' - a)$$
$$= \frac{1}{2} \int_{-\pi/2}^{\pi/2} dx \cos x = 1$$

The probability $P(a'|a)$ that a statistical ensemble of photons with l density given by (9) crosses a (second) polarizer with axis a' is obviously equal to the probability of finding l' in the intersection of regions R_a and $R_{a'}$ (Fig. 5.2). This implies

$$P(a'|a) = 2 \int_{a'-\pi/4}^{\pi/4+a} dl' \, \rho_1(l' - a) = \frac{1}{2} \int_{2(a'-a)-\pi/2}^{\pi/2} dx \, \cos x$$
$$= \cos^2(a - a') \tag{10}$$

The quantum-mechanical probability (Malus's law) is thus reproduced. This is true, in general, if one assumes that the part of a photon beam transmitted through a polarizer (with axis $\hat{a}$) always has its l-density modified from whatever it was before transmission to the density (9). This sudden change of the density of the l-vectors is an essential ingredient of the present model if the quantum-mechanical probabilities are to be reproduced.

One can now introduce a new feature of the model, which does not modify in any way the previous conclusions but which will turn out to be of great importance in the study of experiments of EPR type. It is a *variable detection probability* (VDP) dependent on a new vector $\boldsymbol{\lambda}$ associated with every photon. The detection probability $D(l, \lambda)$ is assumed dependent both on the new variable

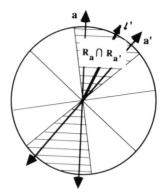

FIGURE 5.2. Hidden-variable model for photons. A photon that has crossed a polarizer with axis $\hat{a}$ and has polarization vector l' will cross a second polarizer with axis $\hat{a}'$ if l' lies in region $R_a \cap R_{a'}$, and be absorbed otherwise.

λ and on the "polarization" variable l. Its λ-average is furthermore assumed to be l-independent and equal to the quantum efficiency η of the photon detector in question:

$$\langle D(l, \lambda) \rangle_\lambda = \eta \tag{11}$$

The value of λ is furthermore assumed to be left unchanged by the photon crossing any optical device (polarizer, prism, half-wave plate, . . .). Its only active role is in the detection process. Of course, no λ-parameter is known within standard quantum theory. However, it is well known that von Neumann's theorem and all theorems of the same type are unable to avoid deterministic generalizations ("completions") of quantum theory. They are *a fortiori* unable to forbid the introductionn of parameters, such as the λ entering in (11), which do not lead to a deterministic model but allow for a more detailed probabilistic description. The usefulness of such a description should therefore be judged only on empirical grounds.

The photon model just developed will now be applied to an EPR-type experiment. A source S emits pairs of photons that are geometrically selected to travel in roughly opposite directions, cross suitable filters (not shown), and impinge on polarizers. The photons traveling to the left and right of Fig. 5.3 will be called L- and R-photon, respectively. Now the L-photon has linear polarization variable l and detection variable λ, as determined earlier. It crosses the polarizer with axis $\hat{a}$ with a probability

$$\begin{aligned} T_1(la) &= 1 \qquad \text{if } l \in R_a \\ &= 0 \qquad \text{if } l \notin R_a \end{aligned} \tag{12}$$

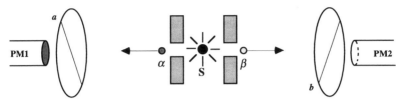

Fig. 5.3. Two correlated photons α and β are emitted simultaneously by the source S, interact with the polarizers with axes **a** and **b**, respectively, and possibly enter the photodetectors PM_1 and PM_2, respectively.

where R_a is defined by $-\pi/4 \leq l - a \leq \pi/4$ and $3\pi/4 \leq l - a \leq 5\pi/4$. If it does cross the polarizer, its l-vector is modified in such a way that its probability density becomes $\rho_l(l' - a)$, as given in (9). The λ-vector is unmodified by the crossing of the polarizer and gives rise to a detection probability $D_1(l', \lambda)$.

Similarly the R-photon, which in the case of the state with $J = 0$ and positive parity is assumed to have the same l-vector as its companion L-photon, will cross the polarizer with axis $\hat{\mathbf{b}}$ with probability

$$T_2(lb) = 1 \qquad \text{if } l \in R_b$$
$$= 0 \qquad \text{if } l \notin R_b \tag{13}$$

where, of course, R_b is defined by $-\pi/4 \leq l - b \leq \pi/4$ and $3\pi/4 \leq l - b \leq 5\pi/4$. If the R-photon crosses the polarizer, its l-vector is modified in such a way that its probability density becomes

$$\rho_2(l'' - b) = \tfrac{1}{2}\cos 2(l'' - b), \qquad l'' \in R_b$$
$$= 0, \qquad l'' \notin R_b \tag{14}$$

The λ'-vector of the R-photon is unmodified by the crossing of the polarizer and gives rise to the detection probability $D_2(l'', \lambda')$.

Assuming an isotropic initial distribution of the l-vectors, one has for the coincidence probability $\omega(a, b)$:

$$\omega(a, b) = \frac{1}{\pi} \int_0^\pi dl \int d\lambda \, d\lambda' \, \rho(\lambda - \lambda') \int_{R_a} dl' \, \rho_1(l' - a) \int_{R_b} dl'' \, \rho_2(l'' - b)$$
$$\times D_1(l' - \lambda)T_1(l - a)T_2(l - b)D_2(l'' - \lambda') \tag{15}$$

where all functions of two variables depend only on the difference of their arguments because of space isotropy. Obviously, $\omega(a - b)$ as given by (15) will factor in a double-transmission probability and a double-detection probability, respectively given by

$$T_{12}(a - b) = \frac{1}{\pi} \int_0^\pi dl \, T_1(l - a)T_2(l - b) \tag{16}$$

and

$$D_{12}(a - b) = \int d\lambda \, d\lambda' \, \rho(\lambda - \lambda') \int_{R_a} dl' \, \rho_1(l' - a) \int_{R_b} dl'' \, \rho_2(l'' - b)$$
$$\times D_1(l' - \lambda)D_2(l'' - \lambda') \tag{17}$$

The calculation of T_{12} is straightforward with the definitions (12) and (13) of $T_1(l - a)$ and $T_2(l - b)$ since both photons can cross their respective polarizers only if l belongs to $R_a \cap R_b$:

$$T_{12}(a - b) = \frac{1}{\pi} \int_{-\pi/4+b}^{\pi/4+a} dl = \frac{1}{2}\left[1 - \frac{2}{\pi}(b - a)\right] \tag{18}$$

where angles are assumed to increase counterclockwise and $b > a$. Result (18) is valid for only $0 \leq b - a \leq \pi/2$. It can be generalized for $-\pi/2 \leq b - a \leq \pi/2$ (that is, in practice, for all angles) and for realistic experimental conditions and gives

$$T_{12}(a - b) = \frac{1}{4}\left[(\epsilon_M + \epsilon_m)^2 + (\epsilon_M - \epsilon_m)^2 F\left(1 - \frac{4}{\pi}|b - a|\right)\right] \tag{19}$$

For the double-detection probability one can write

$$\Delta_{12}(l' - l'') = \int d\lambda \, d\lambda' \, \rho(\lambda - \lambda')D_1(l' - \lambda)D_2(l'' - \lambda')$$
$$= \sum_n \{\alpha_n \cos 2n(l' - l'') + \beta_n \sin 2n(l' - l'')\} \tag{20}$$

where a Fourier expansion in terms of $2(l' - l'')$ has been assumed for $\Delta_{12}(l' - l'')$ with coefficients α_n and β_n. When (20) is inserted in (17) one obtains

$$D_{12}(a - b) = \sum_n \{\alpha_n C_n + \beta_n S_n\} \tag{21}$$

where

$$C_n = \int_{a-\pi/4}^{a+\pi/4} dl' \cos 2(l' - a) \int_{b-\pi/4}^{b+\pi/4} dl'' \cos 2(l'' - b) \cdot \cos 2n(l' - l'')$$
$$S_n = \int_{a-\pi/4}^{a+\pi/4} dl' \cos 2(l' - a) \int_{b-\pi/4}^{a+\pi/4} dl'' \cos 2(l'' - b) \cdot \cos 2n(l' - l'') \tag{22}$$

A rather lengthy but straightforward calculation leads to the results

$$C_n = \frac{1}{(n^2 - 1)}\cos 2n(a - b)$$
$$S_n = \frac{1}{(n^2 - 1)}\sin 2n(a - b) \tag{23}$$

if n is even, and to $C_n = S_n = 0$ if n is odd.

Therefore,

$$D_{12}(a - b) = \sum_{n,\text{even}} \left\{ \frac{\alpha_n}{(n^2 - 1)^2} \cos 2n(a - b) + \frac{\beta_n}{(n^2 - 1)^2} \sin 2n(a - b) \right\} \quad (24)$$

Retaining only the first few terms of this result, one has, for ω,

$$\omega(a - b) = T_{12}(a - b)[1 + \beta \sin 4(a - b) + \cdots] \eta_1 \eta_2 \quad (25)$$

where η_1 and η_2 are again the quantum efficiencies of the two counters, to which the averages of D_1 and D_2 have to be identified, and β is an unknown parameter.

It is gratifying that the correction to T_{12}, due to correlated detections from the two counters, has exactly the form needed in order to make the straight-line correlation provided by T_{12} approach the quantum-mechanical formula. In fact, $\sin 4(a - b)$ vanishes exactly in the points where

$$\frac{1}{2} \left[1 - \frac{2}{\pi} |b - a| \right] = \frac{1}{4}[1 + \cos 2(b - a)]$$

that is, for $|b - a| = 0$, $\pi/4$, $\pi/2$. Furthermore, if $\beta > 0$, $\sin 4(a - b)$ gives rise to an increased correlation for $0 < |a - b| < \pi/4$, since the second term in brackets in (25) is here larger than 1. Similarly, the same term gives rise to a decreased correlation for $\pi/4 < |a - b| < \pi/2$.

Numerically one can obtain a fair agreement with quantum mechanics (and a violation of the CHSH inequality) by taking $\beta = 0.139$. This VDP model, although local, violates the Freedman (strong) inequality. For example,

$$\omega\left(\frac{\pi}{8}\right) - 3\omega\left(\frac{3\pi}{8}\right) = .103 \eta_1 \eta_2 \quad (26)$$

Therefore, the left side of (26) is positive, while the Freedman inequality would require it to be negative or, at most, zero, as seen in Section 3.4.4. It can be checked that $T_{12}(a - b)$ satisfies the CHSH inequality, as one would expect with a local model without VDP.

Naturally the agreement between the present model and quantum theory can be improved by considering further terms in (24).

5.3.2. Divergence with Respect to Quantum Theory

The two-variable photon model developed in the previous sections agrees perfectly with quantum theory for single-photon experiments and can be made to agree very closely with EPR-type experiments performed to the present—that is, with experiments in which each photon is made to cross a single analyzer before entering a detector (Fig. 5.3).

However, the same model disagrees markedly with the predictions of quantum theory if more than one analyzer is put on the trajectory of the photons.[25] To see this consider the apparatus of Fig. 5.4 with two polarizers and two half-wave plates. Let the axes of the two polarizers be **a** and **b** as before and the axes of the two half-wave plates **a'** and **b'**. The physical action of a half-wave plate on a beam of photons compared with that of a polarizer is similar in some respects and different in others. It is similar because the photons emerging from it are in a well-defined quantum state (they are polarized). It is different because, in principle at least, all photons are transmitted and there is no polarization-dependent absorption. In the model considered here a well-defined quantum state is represented by a density $\rho(l' - a)$ as given by (9). Therefore, if the photons cross the half-wave plate (and in principle all of them do), their l'-density must become $\rho(l' - a')$ (where the new l' is totally unrelated to the old one). A similar situation holds for the other photon, which emerges from the two analyzers with a density $\rho(l'' - b')$. Therefore *the double-detection probability gives rise to a correlation only between a' and b'*, and no further correlation between a and b can be generated by the apparatus of Fig. 5.4. The only correlation of the latter type is the one due to the double transmission through the two polarizers. The coincident probability for the apparatus of Fig. 5.4 is therefore

$$\omega = T_{12}(a - b)D_{12}(a' - b') \qquad (27)$$

where

$$T_{12}(a - b) = \frac{1}{4}\left[(\epsilon_M + \epsilon_m)^2 + (\epsilon_M + \epsilon_m)^2 F\left(1 - \frac{4}{\pi}|b - a|\right)\right] \qquad (28)$$

as before, and

$$D_{12}(a' - b') = \sum_{n,\text{even}}\left\{\frac{\alpha_n}{(n^2 - 1)^2}\cos 2n(a' - b') + \frac{\beta_n}{(n^2 - 1)^2}\sin 2n(a' - b')\right\} \qquad (29)$$

In (27) two factors that account for the absorption of the half-wave plates have been neglected because in practice they are never very different from unity.

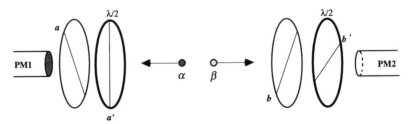

FIGURE 5.4. Two correlated photons α and β interact with polarizers and with half-wave plates before entering the photodetectors *PM*1 and *PM*2, respectively.

The prediction (27) is rather different from the quantum-theoretical formula: In usual quantum theory the change of polarization before detection has no influence whatsoever since no enhanced correlated detections exist in this theory and the double coincident probability is exactly the same for the two experiments of Figs. 5.3 and 5.4:

$$\omega_{QM}(a-b) = \tfrac{1}{4}[(\epsilon_M + \epsilon_m)^2 + (\epsilon_M - \epsilon_m)^2 F \cos 2(a-b)]\eta_1\eta_2 \qquad (30)$$

In one experiment[26] $\epsilon_M = .91$, $\epsilon_m = .03$, and $F = .996$. Therefore,

$$\omega_{QM}(a-b) \simeq .221[1 + f \cos 2(a-b)]\eta_1\eta_2, \qquad \text{with } f = .873 \qquad (31)$$

In the case of only one analyzer, the VDP model can be made to agree with quantum theory very closely. So one can write with good approximation

$$T_{12}(a-b)D_{12}(a-b) = \omega^{QM}(a-b) \qquad (32)$$

This relation holds for arbitrary values of a and b: writing it for a' and b' and substituting $D_{12}(a'-b')$ in (27), one has

$$\omega = \frac{T_{12}(a-b)}{T_{12}(a'-b')}\omega^{QM}(a'-b') \qquad (33)$$

This last result should be compared with the quantum-theoretical expression for the experiment for Fig. 5.4 which does not depend on a' and b' and is, again,

$$\omega^{QM}(a-b) \qquad (34)$$

where (31) still holds. The ratio γ of the two predictions (33) and (34) is thus

$$\gamma(a,b,a',b') = \frac{\omega^{QM}(a-b)}{\omega} = \frac{T_{12}(a'-b')}{T_{12}(a-b)}\frac{\omega^{QM}(a-b)}{\omega^{QM}(a'-b')} \qquad (35)$$

Substitution of (19) and (31) leads to

$$\gamma = \frac{1 + f(1 - (4/\pi)|a'-b'|)}{1 + f(1 - (4/\pi)|a-b|)}\frac{1 + f \cos 2(a-b)}{1 + f \cos 2(a'-b')} \qquad (36)$$

A numerical calculation shows that the largest possible value of γ is obtained for

$$|a-b| = \frac{\pi}{8} \qquad \text{and} \qquad |a'-b'| = \frac{7\pi}{16}$$

where

$$\gamma \simeq 2.01 \qquad (37)$$

rather than $\gamma = 1$, as expected within standard quantum theory.

5.3.3. Enhanced Particle Detection

A detailed study of enhancement has been undertaken by R. Risco Delgado.[27]

If de Broglie's model of a particle embedded in a wave is adopted, the EPR experiments suggest that it be enriched by certain variables, which give rise to a variable probability of detection. The following assumption shall be made: as detection appears to be more closely related to the corpuscular than to the undulatory nature of matter (the particle is detected where it is and not everywhere on the wave front), the amplitude $y(x, t)$ of an oscillatory movement connected with the particle ought to be related to the probability of detection. De Broglie gave an interpretation of his two wave functions, the objectively real wave $\varphi(x, t)$ and the probabilistic wave $\psi(x, t)$, but he left $y(x, t)$ essentially uninterpreted. Perhaps the discovery of "enhancement" was necessary for a meaning to be attributed to $y(x, t)$ as well.

A specific model will now be described quantitatively. A photon consists of a wave and a particle, the two oscillating in phase as in de Broglie's theory; $\hat{l}$ will represent the direction of this vibration, which will be fixed by the angle of polarization of the photon with respect to a certain frame of reference: wave and particle vibrate in direction $\hat{l}$ after being emitted by the source. The probability D of detection is proportional to $\|y(x, t)\|^2$:

$$D = c\|y(x, t)\|^2$$

which is naturally related to the efficiency of the detector.

When particle and wave go through a polarizer whose axis is parallel to $\hat{a}$, both will vibrate along in new direction of polarization $\hat{a}$ and the amplitudes of their movements will be multiplied by $\cos(l - a)$, where $l - a$ is the angle between the vectors $\hat{l}$ and $\hat{a}$. This is just Malus's law extended to the particle's internal vibration.

The particle is assumed to acquire two important characteristics:

(i) The aforementioned law relating the probability of detection to the amplitude of the particle's oscillatory movement is modified, and the probability of detection rises to

$$D = \frac{4}{3}c\|y(x, t)\|^2 \tag{38}$$

This expression is only valid in the domain of low efficiencies. In other words, an "enhancement" results.

(ii) The subsequent polarizers will only reduce the amplitude of the aforementioned motions by a factor of $\sqrt{\cos(l - a)}$. Such variables, which are activated by crossing a polarizer, are common to various kinds of models with variable probability of detection.

Equation (38) should be emphasized. Again, let us suppose a polarizer produces an enhancement of $4/3$. This factor is precisely what is required to hide the variable probability of detection in all single-photon physics and to reproduce atomic-cascade experiments with pairs of photons.

Before facing the calculations one should remember that the probability $P(\mathbf{x}, t)$ for a particle to be found at position $\mathbf{x}$ at time t is proportional to the intensity of the electromagnetic field at that point, $\|\varphi(\mathbf{x}, t)\|^2$. One can therefore use the normalized function $\psi(\mathbf{x}, t)$, related to $\varphi(\mathbf{x}, t)$ through the expression

$$P(\mathbf{x}, t) = |\psi(\mathbf{x}, t)|^2 = K\|\varphi(\mathbf{x}, t)\|^2$$

where K is a normalization constant.

To see how the model works in different experimental situations, assume that the well-known physics of the single photon is satisfied.

5.3.3.1. One Detector

Assume an atom emits a photon in the direction of the detector. Sooner or later the photon will certainly reach the detector if it does not encounter obstacles on the way. Even then, the photon may not be detected, because the detector may not be perfect. The probability ω of the detector clicking will, in general, be equal to the probability P that the photon reaches it times the probability D that, once it has arrived, it is detected:

$$\omega = PD \qquad (39)$$

The field associated with the photon emerging from the source can be described by the function

$$\varphi(\mathbf{x}, t) = \varphi_0(\mathbf{x}, t)\hat{\boldsymbol{l}} \qquad (40)$$

which represents a wave packet vibrating in direction $\hat{\boldsymbol{l}}$ and traveling from the source to the detector. The form of $\varphi_0(\mathbf{x}, t)$ will not be specified because it is not relevant here.

Once the wave packet reaches the detector (Fig. 5.5), the probability P that the photon is at the detector is clearly equal to unity. The field at this point will be represented by

$$\varphi(t) = \varphi_0(t)\hat{\boldsymbol{l}}$$

Because the probability of presence is related to it through equation

$$P(\mathbf{x}, t) = K\|\varphi(\mathbf{x}, t)\|^2$$

FIGURE 5.5. The most elementary measurement: photon α is about to enter the photodetector *PM*.

the normalization constant K can be identified:

$$1 = K\|\boldsymbol{\varphi}(t)\|^2 = K|\varphi_0(t)|^2 \tag{41}$$

On the other hand, the probability D of the detector clicking figures in Eq. (39). The oscillatory motion of the particle will be represented by the function

$$\mathbf{y}(\mathbf{x}, t) = y_0(\mathbf{x}, t)\hat{\boldsymbol{l}}$$

and at the detector

$$\mathbf{y}(t) = y_0(t)\hat{\boldsymbol{l}}$$

so that

$$D = c\|\mathbf{y}(\mathbf{x}, t)\| = c|y_0(t)|^2 = \eta \tag{42}$$

Summing up,

$$\omega = PD = |\varphi_0(t)|^2 c|y_0(t)|^2 = \eta$$

5.3.3.2. A Detector and a Polarizer

After crossing the polarizer aligned along $\hat{\mathbf{a}}$ (see Fig. 5.6), the wave $\varphi(\mathbf{x}, t)$ and the oscillatory movement of the particle $\mathbf{y}(\mathbf{x}, t)$ become

$$\varphi(\mathbf{x}, t) = \varphi_0(\mathbf{x}, t)\cos(l - a)\hat{\mathbf{a}}$$

and

$$\mathbf{y}(\mathbf{x}, t) = y_0(\mathbf{x}, t)\cos(l - a)\hat{\mathbf{a}}$$

while

$$\varphi(t) = \varphi_0(t)\cos(l - a)\hat{\mathbf{a}}$$

and

$$\mathbf{y}(t) = y_0(t)\cos(l - a)\hat{\mathbf{a}}$$

at the detector

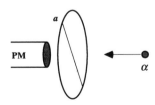

FIGURE 5.6. Photon α is about to interact with the polarizer with axis **b** and **a** before eventually entering the photodetector *PM*.

As $\omega = PD$,

$$P = K\|\boldsymbol{\varphi}(t)\|^2$$

and taking account of (41),

$$P(l) = \frac{1}{|\varphi_0|^2}\|\boldsymbol{\varphi}(t)\|^2 = \cos^2(l - a)$$

Similarly, as a polarizer has been crossed and hence (38) must be applied:

$$D = \tfrac{4}{3}c\|\mathbf{y}\|^2 = \tfrac{4}{3}c|\mathbf{y}_0|\cos^2(l - a)$$

Taking account of (42) gives

$$D(l) = \tfrac{4}{3}\eta\cos^2(l - a).$$

Hence,

$$\omega(l) = P(l)D(l) = \tfrac{4}{3}\eta\cos^4(l - a) \tag{43}$$

This formula will hold for every photon. The polarization of the photons is unknown, however, until they reach the polarizer. The light used in the cascade experiments is unpolarized, which is expressed here by the uniform distribution of the polarization vector $\hat{\boldsymbol{l}}$ being constant. That is, all the values of $\hat{\boldsymbol{l}}$ have the same probability of being emitted:

$$\rho(l)dl = \frac{1}{2\pi}dl$$

Averaging (43) over l one obtains

$$\langle\omega(l)\rangle = \frac{1}{2}\eta$$

which confirms the well-established experimental result that the probability of detection (intensity) of a beam of nonpolarized light is halved by a polarizer.

5.3.3.3. A Detector and Two Polarizers

Much as before, one can calculate the probability P of presence at the detector once emission has occurred and the probability D of detection once the photon has reached the detector (Fig. 5.7). If, when the photon leaves the source, the electromagnetic field is represented by (40), once the first polarizer has been crossed one can write

$$\boldsymbol{\varphi}(\mathbf{x}, t) = \varphi_0(\mathbf{x}, t)\cos(l - a)\hat{\mathbf{a}}$$

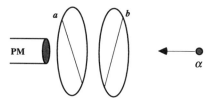

FIGURE 5.7. Photon α is going to interact with two polarizers with axes **b** and **a** before it may reach the photodetector *PM*.

The second polarizer will then attenuate according to $\sqrt{\cos(l-b)}$ so that once the detector is reached, the field associated with the photon will have the form

$$\varphi(t) = \varphi_0(t)\cos(l-a)\cos^{1/2}(l-b)\hat{\mathbf{b}}$$

This value of the field implies that the probability of the photon being present at the detector, after the right lapse of time, will be

$$P(l) = \frac{1}{|\varphi_0(t)|^2}\|\varphi(t)\| = \cos^2(l-a)\cos(l-b)$$

After crossing the first polarizer the oscillatory motion of the particle is reduced by the same factor $\cos(l-a)$:

$$\mathbf{y}(\mathbf{x}, t) = y_0(\mathbf{x}, t)\cos(l-a)\hat{\mathbf{a}}$$

and the factor $\sqrt{\cos(l-b)}$ applies again after the second has been crossed, so at the detector this motion will have been attenuated according to

$$\mathbf{y}(t) = y_0(t)\cos(l-a)\cos^{1/2}(l-b)\hat{\mathbf{b}}$$

Therefore, as the polarizers have been crossed,

$$D = \tfrac{4}{3}c\|\mathbf{y}\|^2 = \tfrac{4}{3}c|\mathbf{y}_0|^2\cos^2(l-a)\cos(l-b)$$

and therefore, taking account of (42),

$$D(l) = \tfrac{4}{3}\eta\cos^2(l-a)\cos(l-b)$$

Hence,

$$\omega(l) = P(l)D(l) = \tfrac{4}{3}\eta\cos^4(l-a)\cos^2(l-b)$$

Again, this result has to be averaged. Under the same conditions as before, that is, for the same distribution function $\rho(l)$, one has

$$\langle\omega(l)\rangle = \frac{1}{2\pi}\frac{4}{3}\eta\int_0^{2\pi}\cos^4(l-a)\cos^2(l-b)\,dl$$

so that

$$\langle \omega(l) \rangle = \tfrac{1}{2} \eta \cos^2(a - b)$$

which confirms the expected result.

With these three cases the first objective has been reached, which was to reproduce exactly the physics of the beam of photons—single-photon physics—with this simple model in which a new role, which is directly involved in the detection process, it attributed to the vibration of the particle. It should be emphasized that in this model, the same mechanism of "enhancement" that accounts for the experimental results involving pairs of photons also works in single-photon physics. This is reassuring; it would be disturbing if enhancement were only involved in processes where it is absolutely necessary. It is perfectly hidden, however, where there is just a single beam.

5.3.4. Correlated Detections of Two Photons

The following setups are the very ones used in two-photon experiments. A pair of photons whose polarizations are correlated is generated by a source, which may, for instance, be the cascade of ^{40}Ca used by Aspect *et al.*[28] The most natural way, in this model, to understand the correlation of the photons' polarizations is to assume that they have the same polarization vector $\hat{l}$. This will be irrelevant in Sections 5.3.4.1 and 5.3.4.2 but vital for 5.3.4.3, which is the case directly related to the experiments of interest.

5.3.4.1. Two Detectors

Now consider the probability ω_{12} that both detectors click when a cascade has taken place at the source, emitting two photons in the direction of the two detectors (Fig. 5.8). Neither photon interacts with a polarizer before reaching the detector, so their associated fields will have the forms

$$\boldsymbol{\varphi}_k(t) = \varphi_{0k}(t)\hat{l}, \qquad k = 1, 2,$$

when the detectors are reached. The oscillations of the particles will be represented by

$$\mathbf{y}_k(t) = y_{0k}(t)\hat{l}$$

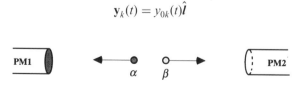

FIGURE 5.8. Two correlated photons α and β enter the photodetectors *PM*1 and *PM*2, respectively.

The probability of a joint detection will be equal to the probability that one clicks times the probability that, once the first has clicked, the second clicks. The probability ω_1 that the first detector clicks will be the probability P_1 that the particle reaches it (equal to unity because we are assuming the particles are emitted in the direction of the detectors) times the probability D_1 that, once the particle reaches the detector, it clicks. As $P_1 = 1$, and the photons are produced in pairs, P_2, the probability that the second photon reaches the second polarizer, will also be equal to unity. That is,

$$\omega_{12} = \omega_1 \omega_2 = P_1 D_1 P_2 D_2$$

with

$$P_k = K \| \varphi_k \|^2 = 1 \tag{44}$$
$$D_k = c \| \mathbf{y}_k(t) \|^2 = c | \mathbf{y}_{0k}(t) |^2 = \eta_k \tag{45}$$

where $k = 1, 2$, so that

$$\omega_{12} = \eta_1 \eta_2$$

which coincides with the predictions of quantum mechanics and experimental results.

5.3.4.2. Two Detectors and a Polarizer

Assume, as usual, that a cascade emission has taken place and that therefore a pair of photons has certainly emerged from the source, and, furthermore, the two photons are traveling in the direction of the analyzers (Fig. 5.9). The probability of a double detection will be

$$\omega_{12} = \omega_1 \omega_2 = P_1 D_1 P_2 D_2$$

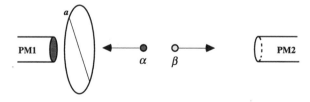

FIGURE 5.9. Of two correlated photons α and β, the first (α) interacts with a polarizer with axis **a** before entering the photodetector *PM*1, while the second (β) enters the photodetector *PM*2 directly.

The functions P_1, P_2, D_1, and D_2 will now be evaluated. For photon 1, after emission by the source,

$$\varphi_1(\mathbf{x}, t) = \varphi_{01}(\mathbf{x}, t)\hat{\boldsymbol{l}} \tag{46}$$

$$\mathbf{y}_1(\mathbf{x}, t) = y_{01}(\mathbf{x}, t)\hat{\boldsymbol{l}} \tag{47}$$

and since this photon will not cross a polarizer, at the detector one can write

$$\varphi_1(t) = \varphi_{01}(t)\hat{\boldsymbol{l}}, \qquad \mathbf{y}_1(t) = y_{01}(t)\hat{\boldsymbol{l}}$$

Hence,

$$P_1 = K\|\varphi_1(t)\|^2 = 1$$

$$D_1 = c\|\mathbf{y}_1(t)\|^2 = c|y_{01}(t)|^2 = \eta_1$$

where account has been taken of (44) and (45).

Similarly, for photon 2, after emission by the source,

$$\varphi_2(\mathbf{x}, t) = \varphi_{02}(\mathbf{x}, t)\hat{\boldsymbol{l}} \tag{48}$$

$$\mathbf{y}_2(\mathbf{x}, t) = y_{02}(\mathbf{x}, t)\hat{\boldsymbol{l}} \tag{49}$$

But now this photon encounters a polarizer, and hence the functions become

$$\varphi_2(t) = \varphi_{02}(t)\cos(l - a)\hat{\mathbf{a}}$$

$$\mathbf{y}_2(\mathbf{x}, t) = y_{02}(\mathbf{x}, t)\cos(l - a)\hat{\mathbf{a}}$$

at the detector. So

$$P_2(l) = K\|\varphi_{02}\|^2 = \cos^2(l - a)$$

$$D_2(l) = c\|\mathbf{y}_2(t)\|^2 = c|y_{02}(t)|^2 = \tfrac{4}{3}\eta_2 \cos^2(l - a)$$

where account has been taken of (44) and (45). Hence,

$$\omega_{12}(l) = \tfrac{4}{3}\eta_1\eta_2 \cos^4(l - a)$$

The fact that the light leaving the calcium source does not have a definite polarization will now be expressed as

$$\rho(l)dl = \frac{1}{2\pi}dl$$

Averaging over l,

$$\langle\omega_{12}(l)\rangle = \tfrac{1}{2}\eta_1\eta_2$$

which agrees with the expected result.

5.3.4.3. Two Detectors and Two Polarizers

Here the very type of apparatus figuring in experiments used to study the possibility of nonlocality will be considered. Until now the model has passed all tests. Enhancement has already been implicitly at work. This, however, is the decisive case (Fig. 5.10).

Emerging from the source, the photons will be characterized by functions (46)–(49). Once the polarizers oriented along axes $\hat{\mathbf{a}}_1$ and $\hat{\mathbf{a}}_2$ have been crossed, the fields at the detectors will be described by

$$\boldsymbol{\varphi}_1(t) = \varphi_{01}(t)\cos(l - a_1)\hat{\mathbf{a}}_1$$
$$\mathbf{y}_1(t) = y_{01}(t)\cos(l - a_1)\hat{\mathbf{a}}_1$$
$$\boldsymbol{\varphi}_2(t) = \varphi_{02}(t)\cos(l - a_2)\hat{\mathbf{a}}_2$$
$$\mathbf{y}_2(t) = y_{02}(t)\cos(l - a_2)\hat{\mathbf{a}}_2$$

The joint detection probability ω_{12} will be given by

$$\omega_{12} = P_1 D_1 P_2 D_2$$

where

$$P_k(l) = K\|\boldsymbol{\varphi}_k(t)\|^2 = \cos^2(l - a_k)$$
$$D_k(l) = c\|\mathbf{y}_1(t)\|^2 = c|y_{0k}(t)|^2 = \tfrac{4}{3}\eta_k \cos^2(l - a_k)$$

Hence,

$$\omega_{12}(l) = \frac{16}{9}\eta_1\eta_2 \cos^4(l - a_1)\cos^4(l - a_2)$$

Averaging over l gives

$$\langle\omega_{12}(l)\rangle = \frac{1}{4}\eta_1\eta_2\left[1 + \frac{8}{9}\cos 2(a_1 - a_2) + \frac{1}{18}\cos 4(a_1 - a_2)\right]$$

The expression predicted by quantum mechanics in this case is

$$\langle\omega_{12}(l)\rangle = \tfrac{1}{4}\eta_1\eta_2[1 + \cos 2(a_1 - a_2)]$$

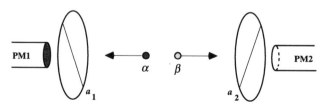

FIGURE 5.10. Standard EPR experiment: two photons (α and β), two polarizers (with axes $\mathbf{a}_1$ and $\mathbf{a}_2$), and two detectors (*PM1* and *PM2*).

The predictions are rather similar, especially if it is remembered that fringe visibility equal to unity is practically impossible to obtain. In the next section further refinements, which take account of the nonideal nature of the polarizers, shall be introduced.

5.3.3.4. Two Detectors and Two Nonideal Polarizers

Until now it has been assumed that the polarizers are ideal, that is, that all the light emerging from them, attenuated according to Malus's law, is polarized in a single plane. Things are not so simple, however there is always a small chance of finding light polarized perpendicular to the principal axis of the polarizer, that is, along the secondary axis. The model will have to be refined to take account of such effects. Henceforth a photon polarized along axis $\hat{\mathbf{l}}$ impinging on a polarizer oriented along direction $\hat{\mathbf{a}}$ has a probability ϵ_M of emerging oriented along direction $\hat{\mathbf{a}}$; the amplitude of its associated wave and of the vibratory movement of the particle are reduced by a factor of $\cos(l - a)$. But there is also a probability ϵ_m of it emerging oriented along $\hat{\mathbf{a}}_{\pi/2}$, where $\hat{\mathbf{a}}_{\pi/2}$ is a vector forming a right angle with $\hat{\mathbf{a}}$ and contained in the plane perpendicular to the motion. In this case the corresponding amplitudes are reduced by a factor $\cos(l - a + \pi/2)$.

The probability of a double detection will be the sum of four probabilities, since there are four possibilities: both can emerge from the principal channels, both from the secondary channels, or one from each. Therefore,

$$\omega_{12} = P_M^1 D_M^1 P_M^2 D_M^2 + P_m^1 D_m^1 P_m^2 D_m^2 + P_M^1 D_M^1 P_m^2 D_m^2 + P_m^1 D_m^1 P_M^2 D_M^2$$

where P_M^1, for instance, is the probability that photon 1 emerges from the principal channel of polarizer 1, and D_M^1 is the probability that if the latter has taken place the particle gets detected. The other probabilities are defined similarly.

A straightforward calculation leads to

$$\langle \omega_{12}(l) \rangle = \frac{1}{2\pi} \frac{16}{9} \eta_1 \eta_2$$

$$\times \left\{ \epsilon_M^1 \epsilon_M^2 \frac{9}{64} 2\pi \left(1 + \frac{8}{9} \cos 2(a_1 - a_2) + \frac{1}{18} \cos 4(a_1 - a_2) \right) \right.$$

$$+ \epsilon_m^1 \epsilon_m^2 \frac{9}{64} 2\pi \left(1 + \frac{8}{9} \cos 2(a_1 - a_2) + \frac{1}{18} \cos 4(a_1 - a_2) \right)$$

$$+ \epsilon_M^1 \epsilon_M^2 \frac{9}{64} 2\pi \left(1 + \frac{8}{9} \cos 2(a_1 - a_2) + \frac{1}{18} \cos 4(a_1 - a_2) \right)$$

$$\left. + \epsilon_M^1 \epsilon_m^2 \frac{9}{64} 2\pi \left(1 + \frac{8}{9} \cos 2(a_1 - a_2) + \frac{1}{18} \cos 4(a_1 - a_2) \right) \right\} \quad (50)$$

Defining

$$\epsilon_\pm^1 = \epsilon_M^1 \pm \epsilon_m^1, \qquad \epsilon_\pm^2 = \epsilon_M^2 \pm \epsilon_m^2$$

Equation (50) can be rewritten as

$$\langle \omega_{12}(l) \rangle = \frac{1}{4}\eta_1\eta_2\left[\epsilon_+^1\epsilon_+^2 + \epsilon_-^1\epsilon_-^2\frac{8}{9}\cos 2(a_1 - a_2) + \epsilon_+^1\epsilon_+^2\frac{1}{18}\cos 4(a_1 - a_2)\right] \tag{51}$$

These results can be compared with the predictions of quantum mechanics with an experiment performed by Aspect,[28] in which the authors claimed to have refuted local realism by 13 standard deviations. In this experiment

$$\epsilon_M^1 = 0.971 \pm 0.005 \qquad \epsilon_M^2 = 0.969 \pm 0.005$$
$$\epsilon_m^1 = 0.029 \pm 0.005 \qquad \epsilon_m^2 = 0.028 \pm 0.005$$

For these values of the transmittancies, the quantum-mechanical predictions give

$$\langle \omega_{12} \rangle = [0.249 + 0.221 \cos 2(a_1 - a_2)]\eta_1\eta_2$$

while the prediction of this model is obtained from (51):

$$\langle \omega_{12} \rangle = [0.249 + 0.197 \cos 2(a_1 - a_2) + 0.0017 \cos 4(a_1 - a_2)]\eta_1\eta_2$$

As Fig. 5.11 shows, the agreement with quantum-mechanical predictions is good. In this analysis account has not been taken of the depolarization factor F, which is usually close to 1. It has the effect of pulling the two curves together.

5.3.5. Experiments with Three Polarizers

In this section the predictions of the local model for three nonideal polarizers will first be considered. Then the quantum predictions for the same setup will be calculated. The interest of working these out when they should be taken for granted is twofold. The procedure followed, although equivalent to the application of the operator formalism, has a clear realist component; the source is viewed as emitting wave packets traveling in spacetime and once the first is detected the other collapses nonlocally. This procedure appears to be more immediate and intuitive than the one generally used in quantum calculations. Furthermore, what are often called quantum predictions can be very personalized. The section ends with a comparative study of the predictions of the local model, of the quantum predictions, and of experimental data.

In an experimental paper the quotient

$$\frac{R(a_2 - a_3, a_1 - a_3)}{R(a_2 - a_3, \infty)} \tag{52}$$

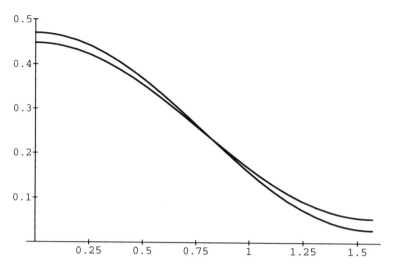

FIGURE 5.11. Quantum mechanics versus the local realistic model with variable detection probability of Eq. (51). In the ordinate the two-photon detection probability (divided by $\eta_1\eta_2$) is represented, in the abscissa the angle between the two polarizers (in radians). The quantum-mechanical curve is slightly higher (lower) for angles near 0 (near $\pi/2$).

was measured, $R(a_2 - a_3, a_1 - a_3)$ being the double-detection rate when all three polarizers are in place and $R(a_2 - a_3, \infty)$ the joint-detection rate when there are only two polarizers ($a_1 - a_3$ is the angle between $\hat{\mathbf{a}}_1$ and $\hat{\mathbf{a}}_3$, and $a_2 - a_3$ the angle between $\hat{\mathbf{a}}_2$ and $\hat{\mathbf{a}}_3$; see Figure 5.12).

The quantity (52) will coincide with

$$\frac{P(a_2 - a_3, a_1 - a_3)}{P(a_2 - a_3, \infty)} \tag{53}$$

$P(a_2 - a_3, a_1 - a_3)$ and $P(a_2 - a_3, \infty)$ being the probabilities of a double detection when there are three and two polarizers, respectively.

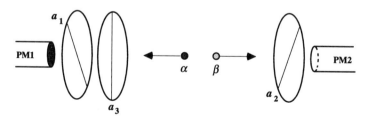

FIGURE 5.12. Modified EPR experiment. Of two correlated photons α and β the first (α) interacts with two polarizers with optical axes $\mathbf{a}_3$ and $\mathbf{a}_1$ before entering the photodetector $PM1$, while the second (β) interacts with the polarizer with axis $\mathbf{a}_2$ before entering the photodetector $PM2$.

The prediction of this model for $P(a_2 - a_3, \infty)$ has already been found and called $\langle \omega_{12}(l) \rangle$; here it shall be applied to the transmittancies of this experiment. Now the same quantity for three polarizers, $\langle \omega_{12}(l) \rangle^{(3)}$, can be worked out. The probability of a double detection will be

$$\omega_{12} = \omega_1 \omega_2$$

where ω_1 is the probability of detector 1 clicking and ω_2 is the probability of detector 2 clicking.

Detector 1 can click because the photon went through the principal channel of polarizer 1, which happens ϵ_M^1 times if the photon is oriented along this axis, and because it went through the secondary channel, the probability of this being ϵ_m^1 if it is oriented along this other axis. Otherwise one would have to work out the corresponding probabilities of the photon going through according to Malus's law, multiplying the last probabilities by P_M^1 and P_m^1. The probability ω_1 will be the sum of these two probabilities. The probability of the photon being detected if it went through the primary channel of the polarizer will be denoted by D_M^1, and the corresponding probability for the secondary channel is D_m^1. Hence,

$$\omega_1 = \epsilon_M^1 P_M^1 D_M^1 + \epsilon_m^1 P_m^1 D_m^1$$

Likewise if detector 2 clicks, photon 2 must have followed one of the four possible channels: primary channel of 3 and primary channel of 2; primary channel of 3 and secondary channel of 2; secondary channel of 3 and primary channel of 2; or secondary channel of 3 and secondary channel of 2. If D_{MM}^2 is the probability of the photon being detected once it has reached detector 2 after having gone through the primary channels of polarizers 3 and 2, and D_{Mm}^2, D_{mM}^2, D_{mm}^2 are defined analogously, one can write

$$\omega_2 = \epsilon_M^3 \epsilon_M^2 P_{MM}^2 D_{MM}^2 + \epsilon_M^3 \epsilon_m^2 P_{Mm}^2 D_{Mm}^2$$
$$+ \epsilon_m^3 \epsilon_M^2 P_{mM}^2 D_{mM}^2 + \epsilon_m^3 \epsilon_m^2 P_{mm}^2 D_{mm}^2 \qquad (54)$$

The various values of P and D will now be calculated. The pair of photons emerging from the source are represented by functions (46)–(49). Once the principal channel of polarizer 1, oriented along $\hat{a}_1$, has been crossed, Eqs. (46) and (47) become

$$\boldsymbol{\varphi}_1^M(t) = \varphi_{01}(t)\cos(l - a_1)\hat{a}_1$$
$$\mathbf{y}_1^M(t) = y_{01}(t)\cos(l - a_1)\hat{a}_1$$

If its secondary channel was crossed instead, one would have

$$\boldsymbol{\varphi}_1^m(t) = \varphi_{01}(t)\cos\left(l - a_1 + \frac{\pi}{2}\right)\hat{a}_{1,\pi/2}$$
$$\mathbf{y}_1^m(t) = y_{01}(t)\cos\left(l - a_1 + \frac{\pi}{2}\right)\hat{a}_{1,\pi/2}$$

at detector 1.

The probabilities associated with photon 1 can now be calculated:

$$P_M^1 = \frac{1}{|\varphi_{01}(t)|} \|\varphi_1^M(t)\|^2 = \cos^2(l - a_1)$$

$$D_M^1 = \frac{4}{3}\eta_1 \frac{1}{|y_{01}(t)|} \|\mathbf{y}_1^M(t)\|^2 = \frac{4}{3}\eta_1 \cos^2(l - a_1)$$

$$P_m^1 = \frac{1}{|\varphi_{01}(t)|} \|\varphi_1^m(t)\|^2 = \sin^2(l - a_1)$$

$$D_M^1 = \frac{4}{3}\eta_1 \frac{1}{|y_{01}(t)|} \|\mathbf{y}_1^m(t)\|^2 = \frac{4}{3}\eta_1 \sin^2(l - a_1)$$

and therefore

$$\omega_1 = \epsilon_M^1 \cos^2(l - a_1)\tfrac{4}{3}\eta_1 \cos^2(l - a_1) + \epsilon_m^1 \sin^2(l - a_1)\tfrac{4}{3}\eta_1 \sin^2(l - a_1)$$
$$= \tfrac{4}{3}\eta_1[\epsilon_M^1 \cos^4(l - a_1) + \epsilon_m^1 \sin^4(l - a_1)]$$

One can likewise work out the probabilities for photon 2. After crossing the (vertically oriented) primary channel of polarizer 3, Eqs. (48) and (49) become

$$\varphi_2^M(t) = \varphi_{02}(t)\cos(l - a_3)\hat{\mathbf{a}}_3$$
$$\mathbf{y}_2^M(t) = y_{02}(t)\cos(l - a_3)\hat{\mathbf{a}}_3$$

at polarizer 2, where $\hat{\mathbf{a}}_3$ indicates the vertical direction. If the secondary channel of polarizer 3 was crossed, at polarizer 2 one would have

$$\varphi_2^m(t) = \varphi_{02}(t)\cos\left(l - a_3 + \frac{\pi}{2}\right)\hat{\mathbf{a}}_{3,\pi/2}$$
$$\mathbf{y}_1^m(t) = y_{02}(t)\cos\left(l - a_1 + \frac{\pi}{2}\right)\hat{\mathbf{a}}_{3,\pi/2}$$

There will be four possibilities for the values assumed by these functions at detector 2 after crossing polarizer 2:

(i)

$$\varphi_2^{MM}(t) = \varphi_{02}(t)\cos(l - a_3)\cos^{1/2}(a_2 - a_3)\hat{\mathbf{a}}_2$$
$$\mathbf{y}_2^{MM}(t) = y_{02}(t)\cos(l - a_3)\cos^{1/2}(a_2 - a_3)\hat{\mathbf{a}}_2$$

(ii)

$$\varphi_2^{Mm}(t) = \varphi_{02}(t)\cos(l - a_3)\cos^{1/2}(a_2 - a_3 + \pi/2)\hat{\mathbf{a}}_{2,\pi/2}$$
$$\mathbf{y}_2^{Mm}(t) = y_{02}(t)\cos(l - a_3)\cos^{1/2}(a_2 - a_3 + \pi/2)\hat{\mathbf{a}}_{2,\pi/2}$$

(iii)

$$\varphi_2^{mM}(t) = \varphi_{02}(t)\cos(l - a_3 + \pi/2)\cos^{1/2}(a_2 - a_3)\hat{\mathbf{a}}_2$$
$$\mathbf{y}_2^{mM}(t) = y_{02}(t)\cos(l - a_3 + \pi/2)\cos^{1/2}(a_2 - a_3)\hat{\mathbf{a}}_2$$

(iv)

$$\varphi_2^{mm}(t) = \varphi_{02}(t)\cos(l - a_3 + \pi/2)\cos^{1/2}(a_2 - a_3 + \pi/2)\hat{a}_{2,\pi/2}$$

$$y_2^{mm}(t) = y_{02}(t)\cos(l - a_3 + \pi/2)\cos^{1/2}(a_2 - a_3 + \pi/2)\hat{a}_{2,\pi/2}$$

The probabilities associated with photon 2 will therefore be

$$P_{MM}^2 = \frac{1}{|\varphi_{02}(t)|}\|\varphi_2^{MM}(t)\|^2 = \cos^2(l - a_3)\cos(a_2 - a_3)$$

$$D_{MM}^2 = \frac{4}{3}\eta_2\frac{1}{|y_{02}(t)|}\|y_2^{MM}(t)\|^2 = \frac{4}{3}\eta_2\cos^2(l - a_3)\cos(a_2 - a_3)$$

$$P_{Mm}^2 = \frac{1}{|\varphi_{02}(t)|}\|\varphi_2^{Mm}(t)\|^2 = \cos^2(l - a_3)\sin(a_2 - a_3)$$

$$D_{Mm}^2 = \frac{4}{3}\eta_2\frac{1}{|y_{02}(t)|}\|y_2^{Mm}(t)\|^2 = \frac{4}{3}\eta_2\cos^2(l - a_3)\sin(a_2 - a_3)$$

$$P_{mM}^2 = \frac{1}{|\varphi_{02}(t)|}\|\varphi_2^{mM}(t)\|^2 = \sin^2(l - a_3)\cos(a_2 - a_3)$$

$$D_{mM}^2 = \frac{4}{3}\eta_2\frac{1}{|y_{02}(t)|}\|y_2^{mM}(t)\|^2 = \frac{4}{3}\eta_2\cos^2(l - a_3)\cos(a_2 - a_3)$$

$$P_{mm}^2 = \frac{1}{|\varphi_{02}(t)|}\|\varphi_2^{mm}(t)\|^2 = \sin^2(l - a_3)\sin(a_2 - a_3)$$

$$D_{mm}^2 = \frac{4}{3}\eta_2\frac{1}{|y_{02}(t)|}\|y_2^{mm}(t)\|^2 = \frac{4}{3}\eta_2\sin^2(l - a_3)\sin(a_2 - a_3)$$

With these results, taking account of (54), one obtains

$$\omega_2 = \tfrac{4}{3}\eta_2[\epsilon_M^3\cos^4(l - a_3)\{\epsilon_M^2\cos^2(a_2 - a_3) + \epsilon_m^2\sin^2(a_2 - a_3)\}$$
$$+ \epsilon_m^3\sin^4(l - a_3)\{\epsilon_M^2\sin^2(a_2 - a_3) + \epsilon_m^2\cos^2(a_2 - a_3)\}]$$

Defining $\epsilon_\pm^2 = \epsilon_M^2 \pm \epsilon_m^2$ and taking account of simple trigonometric relations, one obtains

$$\omega_2 = \tfrac{4}{3}\eta_2[\epsilon_M^3\cos^4(l - a_3)\tfrac{1}{2}\{\epsilon_+^2 + \epsilon_-^2\cos^2 2(a_2 - a_3)\}$$
$$+ \epsilon_m^3\sin^4(l - a_3)\{\epsilon_+^2 - \epsilon_-^2\cos^2 2(a_2 - a_3)\}].$$

Defining also

$$\epsilon_\pm^1 = \epsilon_M^1 \pm \epsilon_m^1, \qquad \epsilon_\pm^3 = \epsilon_M^3 \pm \epsilon_m^3$$

one obtains

$$\langle \omega_{12}(l) \rangle^{(3)} = \frac{\eta_1 \eta_2}{8} \left\{ [\epsilon_+^1 \epsilon_+^2 \epsilon_+^3 + \epsilon_-^1 \epsilon_-^2 \epsilon_-^3 \cos 2(a_1 - a_3)] \right.$$

$$+ \frac{1}{9}[\epsilon_-^1 \epsilon_+^2 \epsilon_-^3 + \epsilon_-^1 \epsilon_-^2 \epsilon_+^3 \cos 2(a_1 - a_3)] \cos 2(a_2 - a_3)$$

$$+ \left. \frac{1}{18}[\epsilon_+^1 \epsilon_+^2 \epsilon_+^3 + \epsilon_+^1 \epsilon_-^2 \epsilon_-^3 \cos 2(a_1 - a_3)] \cos 4(a_2 - a_3) \right\} \quad (55)$$

The prediction of the local model will be

$$\left[\frac{R(a_2 - a_3, a_1 - a_3)}{R(a_2 - a_3, \infty)} \right]^{LR} = \frac{\langle \omega_{12}(l) \rangle^{(3)}}{\langle \omega_{12}(l) \rangle}$$

where $\langle \omega_{12}(l) \rangle$ and $\langle \omega_{12}(l) \rangle^{(3)}$ are given by (51) and by (55), respectively.

5.3.6. The Local Model, Quantum Theory, and Experiments

Since

$$\left[\frac{R(a_2 - a_3, a_1 - a_3)}{R(a_2 - a_3, \infty)} \right]_{QM}$$

$$= \frac{1}{2} \left[\epsilon_+^3 + \epsilon_-^3 \frac{\epsilon_M^2 P - \epsilon_m^2 Q}{\epsilon_M^2 P + \epsilon_m^2 Q} \cos 2(a_1 - a_3) - \Delta(a_2 - a_3, a_1 - a_3) \right]$$

with

$$Q = \epsilon_+^1 + \epsilon_-^1 \cos 2(a_2 - a_3)$$

$$P = \epsilon_+^1 - \epsilon_-^1 \cos 2(a_2 - a_3)$$

$$\Delta(a_2 - a_3, a_1 - a_3) = \frac{\sqrt{\epsilon_M^2 \epsilon_m^2} \epsilon_-^1 \epsilon_-^3}{\epsilon_M^2 P + \epsilon_m^2 Q} \sin 2(a_2 - a_3) \sin 2(a_1 - a_3)$$

The experimental data can now be compared with the local model and with the quantum-mechanical predictions. The Stirling group only found data for three values of the angle $a_2 - a_3$: $a_2 - a_3 = 0°$, $a_2 - a_3 = 33°$, and $a_2 - a_3 = 67.5°$. Of these orientations of the polarizer, the first one is easily described: the two theories are practically impossible to tell apart and agree with experimental data. The second and third cases correspond to Fig. 5.13 and Fig. 5.14. For each value of $a_2 - a_3$, the joint probability of detection was measured for various values of the angle $a_1 - a_3$:

For $a_2 - a_3 = 0°$: $a_1 - a_3 = 0°$, 22.5°, 45°, 67.5°, 90°.
For $a_2 - a_3 = 33°$: $a_1 - a_3 = 0°$, 22.5°, $\approx 33°$, 45°, 67.5°, 90°.
For $a_2 - a_3 = 67.5°$: $a_1 - a_3 = 0°$, 22.5°, 45°, 67.5°, 90°.

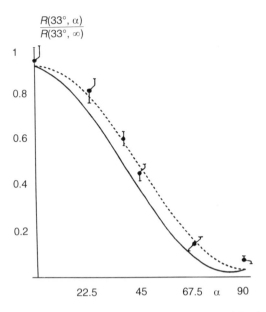

FIGURE 5.13. Quantum-mechanical predictions (continuous line) and local realistic model with variable detection probability (dashed line) compared with the experimental points for $a_1 - a_3 = 33°$ and variable $a_2 - a_3$ (in abscissa). (After Risco Delgado[27])

The graphs reveal that the predictions of the local model are closer than the quantum-mechanical predictions to the experimental values. Also notice the absence of a fundamental point, the one corresponding to $a_2 - a_3 = 67.5°$ and $a_1 - a_3 = 22.5°$. The absence of this point would not be so important if it did not correspond to the very values where the disagreement between the two theories is largest. It really is a pity that the Stirling group decided not to measure this point.

5.4. OTHER REALIST PROPOSALS

5.4.1. Chaotic Ball Model

Caroline Thompson's critical reconsideration of the EPR experiments performed with photon pairs[29] will now be reviewed. Consider the consequences of the fact that, with the very weak signals involved in EPR two-photon experiments, only a fraction of the signals are detected. How does the detector decide which ones? In a semiclassical model of light, the signals in, say, the Orsay experiments must be individually varied in intensity (considered as a classical electromagnetic intensity) as they pass through beam splitters or polarizers.

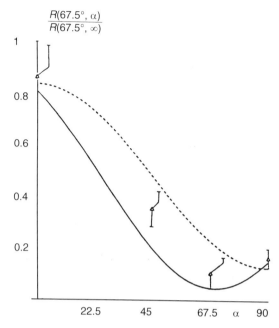

FIGURE 5.14. Quantum-mechanical predictions (continuous line) and local realistic model with variable detection probability (dashed line) compared with the experimental points for $a_1 - a_3 = 67.5°$ and variable $a_2 - a_3$ (in abscissa). (After Risco Delgado[27])

Somehow this intensity variation must determine the probability of detection, and it is assumed that the process is fixed by the sum of the incoming signal and local electromagnetic noise and by the existence of a detection threshold for the sum. A more complete model should perhaps take account of phase and frequency, but the simple picture, with intensity as the only variable, is adequate for understanding qualitatively the logic of these experiments.

The relevance of noise is apparent in the way experimenters have, in practice, to modify the electromagnetic environment of the detector (screening, adjusting temperature), the detection threshold, and the voltage to obtain acceptable behavior. The criteria used for acceptability include an approximate linear response to intensity, but the model under discussion makes linearity over the complete range of possible intensities impossible: No conceivable distribution of noise intensities could produce it. This point is in agreement with the well-known fact that every experimental apparatus has a limited range of applicability. For example, the photodetectors have an efficiency not only strongly dependent on wavelength but also variable with the intensity of the incoming beam. This is, on its own, enough to refute the usual oversimplified picture! Surely, if all photons are identical, as assumed in standard quantum theory, then even very few of them will still all have the same probability of detection. All detectors, at all "quantum efficiencies," should, if the action of a polarizer is simply to reduce the proportion of photons according to Malus's law, reproduce its sinusoidal shape perfectly at all intensities. It is well known that in real experiments this is not the case.

The possibility of a distortion of Malus's law as measured by "counts" in EPR experiments, lies at the heart of the "detection loophole" Consider the usual semiclassical model for an EPR experiment involving pairs of photons correlated in polarization $\hat{\lambda}$, with $\hat{\lambda}$ randomly distributed. A perfect Malus's law response corresponds to a value of

$$S = 2\frac{R(\pi/8) - R(3\pi/8)}{R(\pi/8) + R(3\pi/8)} = 1.414$$

Very weak signals will tend to distort the effective law, producing detections only for a small range of polarization angles, near the polarizer axis, if the assumed intensity threshold is really at work. A little mathematics shows that this is going to lead to increased S values, with an upper limit of 4, well above the Bell limit of 2. Now experimenters are aware of this risk, and they do conduct tests that they believe rule it out, but Thompson does not accept them as being adequate. It is possible, by judicious choice of noise and detection threshold, to organize the setup so that all the average properties are as expected (see, for example, the neat $1:2:4$ proportions for "accidental coincidences" later), but the experimenter cannot ensure that the underlying relationships conform to a particular rule. In Thompson's opinion, distortion of Malus's law could be important in, for example, Aspect's 1982 experiment, in which the value of S based on the raw coincidence counts was 2.169 [figure from Aspect and Grangier, *Lett. Nuovo Cim.* **43**, 345 (1985)].

Thompson has recently concluded that very large violations of Bell-type inequalities are unlikely to be explained by the detection loophole alone. There may be another major factor, an "emission loophole," coming into play. In the majority of experiments, it is not the raw coincidence data that is analyzed but the data after "correction" by subtraction of "accidentals." The kind of data involved is illustrated by the following, which relates to Aspect's 1981 Orsay experiment and was extracted by Thompson from his thesis:

	$R(\pi/8)$	$R(3\pi/8)$	$R(a, \infty)$	$R(\infty, \infty)$
Raw coincidences	86.8	38.3	126.0	248.2
Accidental coincidences	22.8	22.5	45.5	90.0
"Corrected coincidences"	64.0	15.8	80.5	158.2

The table shows coincidence rates (in units of counts per second), using the same notation as in Section 3.5.1. To justify the subtraction, the experimenter has to assume that the emissions are independent. Whether this is fair is highly debatable, but the adjustment is undeniably important. In the current instance, a value of S of 0.776 is converted to one of 1.208.

Thompson's "chaotic ball" model[29] approaches the detection loophole problem from another angle. It presents an intuitive picture, with hidden variables assumed to be distributed over the surface of a sphere. She dismisses certain strong inequalities as invalid: those that estimate probabilities as a ratio with the observed sum of coincidences as the denominator (discussed in Section 3.4.3). The "fair sampling" assumption necessary for their validity may seem reasonable under quantum mechanics but is totally unreasonable for a realist. The chaotic ball model makes it clear that there are circumstances in which the less efficient the detector the greater the violation of these inequalities.

This latter fact is of great importance. Experimenters openly assume the validity of the quantum-mechanical inference that imperfections in the apparatus can only *decrease* the violations. No one so far appears to have fully appreciated that imperfections in the apparatus can *increase* violations with realist models. Because of this assumption, experimenters have (practically) felt free to choose the instrumental settings that gave the strongest violations. It is very unfortunate that Nature appears to encourage bias here, providing all sorts of temptations in the way! As the usual purpose of EPR experiments is to produce an inteference pattern (which is what a coincidence pattern is wrongly perceived to be) and to make it as clear as possible so as to study the positions and spacings of the peaks, how can experimenters be expected to resist opportunities to "improve" it?

In Thompson's opinion other Bell-type inequalities remain valid in principle, however poor the detectors. They are impervious to the "detection loophole" and require no "fair sampling" assumption. They do, however, have another drawback, and perhaps this is the reason for which they are hardly ever used. They depend on being able to compare coincidences with and without polarizers. It is not possible in practice to be sure that no bias is introduced—that there is neither natural nor introduced "enhancement." She nevertheless prefers these inequalities to the others; at least bias is not *necessarily* present. They have very rarely been violated in experiments that do not also involve the "subtraction of accidentals" mentioned earlier. In fact, the only violation seems to be that of Freedman in his 1972 Ph.D. thesis, and here other features of the real experiment may come into play. The particle model is seen to be inadequate not only because it attributes a fixed "size" to the photon but also because it models its spread in *time* incorrectly. There are subtle differences between the quantum-mechanical and semiclassical concepts of coherence length, and confusion over the treatment of emission times.

With a purely undulatory, classical view of light, the most natural interpretation of the observed spectra in the Orsay experiments is that the *A* and *B* "photons" are emitted simultaneously, and each is a wave that starts at high intensity and decreases with a roughly negative exponential rate, only the *A* wave much faster than the *B*. In this model polarizers have the effect of reducing the intensity of each individual incoming signal. Now the "photons" are considerably longer (the *B* one is above 10 ns) than the minimum duration needed for detection. (This appears to be much less than a nanosecond, depending, Thompson believes, more on the time taken waiting for a peak in the noise than on the accumulated energy). Hence, since it is only the first detection that is registered, the average registered time will tend to be later for weaker signals. This can affect the shape of the time spectra and, unless very large windows are used, the logic of the EPR experiments. For when polarizers are parallel, one tends to obtain positive correlations between *A* and *B* intensities, and these become positive correlations in times of detection so synchronization is good. When polarizers are perpendicular, it will be relatively poor, resulting in a higher proportion of large time differences, too large to be recognized as coincidences. This has the net effect of increasing the visibility of the coincidence curve.

Thompson believes that a modest investment in a series of experiments designed to look into the possible realist explanations of published EPR experiments would tell physicists more than the most elaborate "loophole-free" experiment. By investigating performed experiments thoroughly, one could study the responses to a change in, for instance, detector efficiency. Low-efficiency detectors should give the most important information, and she would expect to find clear evidence that the quantum-mechanical approach is inadequate.

Gilbert and Sulcs[30] have arrived at similar conclusions by considering, in particular, the possibility of "coherent noise" causing a spurious increase in correlations. This solution reminds one of stochastic electrodynamics; see, for example, Ref. 31.

5.4.2. Restricted Quantum Mechanics

In this section ideas of certain contemporary realist authors (M. Ferrero, T. Marshall, E. Santos, . . .) will be reviewed, with particular reference to a recent review paper by Ferrero and Santos.[32] They start by discussing the notion of realism and recalling the different brands that have been proposed: convergent, critic, epistemological, gnoseological, internal, intuitive, metaphysical, naïf, ontological, platonic, pragmatic, radical, semantic, etc. To avoid confusion they prefer to state what realism means to them: it is the ontological principle according to which *there is an external world, independent of the subject and his mind, that already existed before it and that will continue to exist when the subject perishes.* Furthermore, this independent external world *can be known.*

The thesis of Hume to refute realism, used by all antirealists until our days was the following:[33]

> What argument could we use to prove that the perceptions of the mind are caused by external objects, entirely different from them, though resembling them (if that may be possible), and cannot have its origin in the energy of the same mind or . . . by any other cause not yet known? In fact it is recognized that many of these perceptions, as in the case of sleep, madness, and some other illnesses, do not have their origin in anything external . . . How to solve this question? By experience, of course, as any other natural question. But in this point the experience is and must be entirely silent. The mind has never anything present to it but the perceptions, and it cannot possibly reach any experience of the connection between perceptions and objects. The supposition of such a connection is, therefore, without any foundation in reasoning.

To this Ferrero and Santos answer that the very existence of physics, chemistry, biochemistry, neurophysiology, etc., explains and confirms the reality of objects, because today one knows how the signals that give rise to the perceptions are emitted, how they are propagated, and, at least partially, how the neurophysiological processes transform the signals into perceptions. The realist postulate so understood is not an arbitrary assumption because the active relation between the subject and the world in which he operates is what proves *a posteriori* the existence of an objective reality.

A further step in the establishment of a rational philosophy for the working physicist is the assumption of locality. A very clear formulation of this idea was given by Einstein:[34]

> If one asks what, irrespective of quantum mechanics, is characteristic of the world of ideas of physics, one is first of all struck by the following: the concepts of physics relate to a real external world, that is, ideas are established relating to things such as bodies, fields, etc., which claim a "real existence" that is independent of the perceiving subject. . . . It is further characteristic of these physical objects that they are thought of as arranged in a space time continuum. An essential aspect of this arrangement of things in physics is that they may claim, at a certain time, to an existence independent of one another, provided these objects "are situated in different parts of space". Unless one makes this kind of assumption about the independence of the existence of objects which are far apart from one another in space . . . physical thinking in the familiar sense would be impossible . . . The following idea characterizes the relative independence of objects far apart in space (A and B): external influence on A has no direct influence on B; this is known as the "principle of contiguity". . . . If this axiom were to be abolished . . . the postulation of laws which can be checked empirically in the accepted sense, would become impossible.

Action at a distance does not exist: If everything could influence anything instantaneously, one would not have been able to observe any regularity, to establish any law.

Ferrero and Santos list 12 interpretations of the existing quantum theory and notice that from the viewpoint of local realism they can be divided into two groups according to the interpretation they give to the wave function ψ:

(i) *Ensemble interpretation.* The wave function ψ describes only an ensemble of similarly prepared systems.

(ii) *Individual interpretation.* The wave function ψ describes every individual system completely.

This distinction was clearly anticipated by Einstein in the 1927 Solvay Meeting. He considered a simple experiment in which electrons, impinging normally on a diaphragm with a small hole, are diffracted through it and eventually detected by a big hemispherical photographic film placed behind. He concluded that there are two ways of understanding the quantum-mechanical formalism, namely:[35]

Viewpoint I. The de Broglie-Schrödinger waves do not represent one individual particle, but rather an ensemble of similar particles distributed in space. In this purely statistical viewpoint $|\psi(r)|^2$ expresses the probability density that there exists at some position r a particle of the ensemble. Accordingly, the theory does not provide any information about the individual process, but only about an ensemble of them.

Viewpoint II. The theory has the pretension of being a complete theory of individual processes. Each particle going to the screen is described by a de Broglie-Schrödinger wave packet. According to this viewpoint, $|\psi(r)|^2$ expresses the probability that at a given moment one and the same particle shows its presence at r while the wave packet instantaneously disappears from all other points of the screen ("collapse of the wave packet"). As long as no localization has been effected, the particle must be considered as potentially present with almost constant probability over the whole area of the screen; however, as soon as it is localized, a peculiar action-at-a-distance must be assumed to take place which prevents the continuously distributed wave in space from producing effects at two or more different places of the screen.

Einstein objected strongly to the second viewpoint, which contains an essential difficulty with its action at a distance, and added[36]:

> "It seems to me that this difficulty cannot be overcome unless the description of the process in terms of the Schrödinger wave is supplemented by some detailed specification of the localization of the particle during its propagation. I think M. de Broglie is right in searching in this direction. If one works only with Schrödinger waves, the interpretation II of $|\psi|^2$, I think, contradicts the postulate of relativity."

Ferrero and Santos agree with Einstein's opinion and observe that if one opts for I within standard quantum theory, then certain problems with realism, with locality, and with relativity emerge. If one instead chooses II, then the conservation laws are endangered for individual processes, something that Einstein thought to have been experimentally refuted by Bothe and Geiger. They choose the statistical interpretation of the wave function and enrich it with local hidden variables, deducing the necessary validity of Bell's inequality.

As is well known, any proof of Bell's theorem has two steps. The first step is the deduction of inequalities from the assumptions of realism and locality, and the second the exhibition of a contradiction with quantum mechanics. Ferrero and Santos accept the first part of the proof. Then they propose a *restriction* of the theory (which they do not consider a real modification) by weakening the standard postulates, which includes removing the postulates of measurement.

The fact that actual experiments have not refuted a genuine (that is, weak) Bell inequality is well known to the experts, but there is a widespread belief that it is only due to the nonideality of the measuring devices (in particular the low efficiency of available photon counters). Furthermore, the loophole is considered a minor practical problem that will surely be solved in the future. This belief has been expressed by many people, including John Bell:[37]

> I always emphasize that the Aspect experiment is too far from the ideal in *many* ways— counter efficiency is only one of them. And I always emphasize that there is therefore a big extrapolation from practical present-day experiments to the conclusion that nonlocality holds. I myself choose to make the extrapolation, for the purpose at least of directing my own future researches. If other people choose differently, I wish them every success and I will watch for their results. In particular, if you can demonstrate that quantum mechanics imposes some limit on the degree to which the ideal experiment can be approached, I will be very interested. I will also be very surprised! Experimental colleagues have told me that optical photon counters could be made as efficient as we like if size and expense are unlimited.

Ferrero and Santos propose the adoption of a "restricted quantum mechanics" (RQM) instead of the usual theory. The main differences with respect to standard quantum theory are that in RQM there are no measurement postulates, and it is neither claimed that all state vectors represent physical states, nor that all Hermitian operators represent observables. The contradiction of such a RQM with local realism cannot be shown as a theorem, e.g., because the violation of Bell's inequality cannot be deduced. From a conceptual point of view the two authors argue that noise has been underestimated in quantum theory: It does not appear in quantum mechanics, but it becomes essential in quantum field theory. They conjecture that noise will prevent the violation of Bell's inequality and, more generally, will eliminate all incompatibility with local realism. Bell's "conjecture" of an irreducible incompatibility between local realism and quantum mechanics, according to them, is false and hence not a theorem. This implies that there are mistakes in the many published proofs of Bell's conjecture.

In modern times, the problem of choosing the right interpretation is "solved" within the ψ interpretation via theories of "decoherence induced by the environment." In the opinion of Ferrero and Santos these theories are welcome, with independence of interpretation, because they show the consistency of quantum mechanics with macroscopic, classical physics. They agree with Einstein, Podolsky, and Rosen that quantum theory is conceptually incomplete, but also with Bohr that it is complete, in the weaker sense that every state realizable in the laboratory may be represented by a density operator. It is not right to claim that the mere "singlet" state vector, involving only spinors, violates the Bell inequality. To see whether the claim is true it is also necessary to know the spatial part of the state vector.

In stochastic theory, noise is the essential ingredient and should be considered an integral part of the theory. Noise may explain in an intuitive form some of the most peculiar predictions of quantum theory, like the Heisenberg uncertainty principle and the probabilistic character of measurement.

The conclusions of this important line of thinking are the following:

1. The principles of realism and locality are fundamental for the construction of physics and therefore ought to be accepted.
2. Local realism leads necessarily to an interpretation of quantum mechanics in terms of hidden variables.
3. It is not true that Bell's inequality has been empirically violated.
4. If the measurement postulates are removed from quantum theory and the vectors–states and operators–observables relations are weakened, then it is not possible to prove "Bell's theorem."
5. Quantum noise will probably prevent an empirical refutation of local hidden variables.

The line of thinking proposed here is very similar to that discussed till now as far as the first three conclusions are concerned. It was shown in Chapter 4, however, that the incompatibility between quantum mechanics and local realism simply arises from the existence of nonfactorizable state vectors, e.g., for two neutral kaons, which do not depend on the parts of quantum axiomatics that Ferrero and Santos consider superfluous. The conclusion proposed here is that quantum theory shall undergo more drastic modifications than the ones envisaged by Ferrero and Santos before it can become compatible with local realism.

5.4.3. A Different Version of the EPR Experiment

An important step forward in the physical understanding of atomic and subatomic phenomena has been made with the realization of the very limited validity of von Neumann's theorem and of all the theorems that ostensibly prove the logical impossibility of a causal completion of quantum mechanics (see Belinfante[38]). The possibility of discovering dispersion-free ensembles, as well as "reduced dispersion" ensembles, has thus become more concrete. It is scientifically reasonable to search for particular statistical ensembles to which the standard quantum rules (like Heisenberg relations) do not apply: these rules would instead be applicable to more general statistical ensembles (which might be called "standard quantum ensembles"). Along these lines, Popper proposed[39] what he called "a new version of the EPR experiment" in which the Copenhagen and statistical interpretations of quantum mechanics could seem to lead to distinguishably different predictions. In this way he hoped, like Einstein, Podolsky, and Rosen, to demonstrate the incompleteness of quantum mechanics, perhaps in a particular experimental situation: A particle can possess position and momentum, in direct conflict with the principle of complementarity upon which the Copenhagen interpretation rests.

Popper considered the experimental arrangement of Fig. 5.15, which he analyzed as follows: A pair of particles is emitted collinearly (in opposite directions) by a source S. A slit at position 1 later physically restricts the y-position of a particle α to precision Δq_y^α and results in observed diffraction there through the Heisenberg relation $\Delta p_y^\alpha \sim \hbar/\Delta q_y^\alpha$, for an ensemble of such pairs. Moreover, since the determination of the position of particles α in this way gives knowledge of the y-position of particle β (through collinearity), this should, according to the Copenhagen interpretation, result in a corresponding "diffraction" of particle β at position 2. (The wide-open slit present at position 2 is there only to prevent photons from reaching the large-angle detectors directly from the source). If this turned out experimentally to be the case, it would be a direct confirmation of the Copenhagen subjectivistic dictum that "knowledge alone"

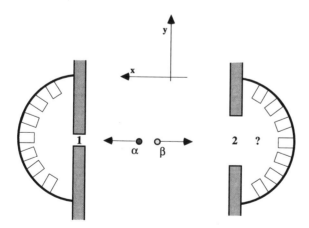

FIGURE 5.15. Popper's ideal experiment. The physical diffraction of the quantum system α through the narrow slit 1 should generate instantaneously an identical diffraction of the correlated quantum system β in the region 2, if knowledge were capable, by itself, of imposing the validity of the uncertainty relations.

can influence a system's behavior. Otherwise it would be in direct conflict with that interpretation, and, in particular, if *nothing* happened to particle β at 2, the arrangement would allow for the determination ("preparation") of the position and momentum of β in violation of the Heisenberg relation $\Delta p_y \Delta q_y \sim \hbar$.

In his analysis Popper accepted that the *physical* restriction of the y-position of an ensemble of particles to a region Δq_y will inevitably result in a distribution of observed y-momenta consistent with the Heisenberg relation; this is what happens to particle α at position 1. Consider then the case where the pairs of correlated particles are photons resulting from the annihilation of electron–positron pairs. (The following analysis applies to all cases in which the physical system that gives rise to the pair of "collinear" particles disappears in producing them.) In the rest frame of each e^+e^- pair the momenta of the emitted photons are equal and opposite, as required. In order to perform the experiment, geometrical considerations clearly require that the e^+e^- annihilations take place in a localized region, the "source," of diameter d, say. This physical restriction (however achieved) will inevitably result in a momentum "uncertainty" (or "scatter," as Popper would have it) $\Delta P_y \sim \hbar/d$, where P is the momentum of the e^+e^- pair before annihilation. Momentum conservation requires that this be imparted to the resulting photon pairs: Collinearity is thereby disrupted to the extent that relative to the momentum of photon α the y-component of the momentum of photon β at the source will be distributed over $\Delta p_y^\beta \sim \hbar/d$. In order for an effect on the momentum of photon β at position 2 to be observable in *principle*, the "restriction" on β's y-position imposed by the action on photon α will have to be such that the corresponding induced momentum "uncertainty" be greater than $\hbar/d$, i.e., that $\Delta q_y^\beta < d$. *As may be seen from Fig. 5.16, this is impossible.*

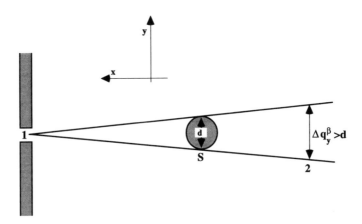

FIGURE 5.16. The diameter (d) of the source of correlated EPR pairs (S) implies an uncertainty for the y component of the position of particle β in the region 2 larger than d.

Thus, any effect on photon β resulting from an action on photon α will be obscured in principle by the inevitable "uncertainty" or "scatter" in p_y^β at the source. So this version of Popper's experiment is bound not to give any useful information.[40]

Positronium decay is, of course, not the only way to obtain pairs of correlated particles. Those sources of geometrically correlated particles, like atoms, which do not disappear in creating the systems of interest will next be considered. In the case of two-photon atomic emission, the situation is different, at least in principle, from the one of the e^+e^- pairs.

As is immediately clear from a consideration of the case of single-photon emission from a source whose size, d, is a few atomic diameters, the momentum uncertainty/scatter of the photon at the source is *not* $\sim h/d$ (which would be far greater than $h\nu/c$!) but is rather $\sim \hbar/c\tau$ (where τ is the decay lifetime and $c\tau$ is the effective confinement of the photon at source). Suppose a photon is emitted in a certain direction in the atomic rest frame. The *atomic* velocity v with respect to the laboratory frame can be evaluated from Heisenberg relations, which give

$$v \simeq \frac{\Delta p_y}{M} \geq \frac{\hbar}{dM}$$

if d, as before, expresses the transversal dimension of the source where the emitting atom (of mass M) is localized. Since in all practical cases one has $v \ll c$, the photon angular deviation in the laboratory is

$$\theta \simeq \frac{v}{c}$$

whence

$$\Delta p_{\text{photon}} \simeq \frac{v}{c}\hbar k \simeq \frac{\hbar}{d}\left(\frac{\hbar k}{Mc}\right) \ll \frac{\hbar}{d}$$

(k is the photon wavenumber).

Similarly, one finds that if two photons are emitted in exactly opposite directions in the center-of-mass frame of the decaying atom, in the laboratory frame the uncertainty/scatter of the momentum of one photon relative to the other is $\sim \hbar/d \cdot (\hbar k/Mc)$, which can clearly be made as small as one wishes by choosing M appropriately. Popper's experiment, if such an ideal source existed, would be possible, at least as far as the limitations imposed by the momentum–position Heisenberg relations are concerned.

Two-photon atomic emission is, however, never collinear *in practice*. Perhaps this reflects the action of the angular-momentum/angle uncertainty scatter. This matter will, however, be avoided, since the latter relations are indeed uncertain. Suffice it to say that they might become meaningful if interpreted to be scatter relations, as by Popper. Apart from this, there does not seem to be any way to rule out *in principle* the possibility that a suitable source of collinear particles (e.g., double Mössbauer effect) might be found. The problem of finding one seems, however, at best very hard to solve in practice.

REFERENCES

1. D. BOHM and B. J. HILEY, *Found. Phys.* **5**, 93–109 (1975); *Found. Phys.* **14**, 255–274 (1984).
2. D. BOHM, in: *Microphysical Reality and Quantum Formalism*, (A. VAN DER MERWE, *et al.*, eds.), Kluwer, Dordrecht (1988), pp. 3–18.
3. J. P. VIGIER, *Lett. Nuovo Cim.* **24**, 258–265 (1979).
4. P. A. M. DIRAC, *Nature* **169**, 702 (1952).
5. J. S. BELL, *Physics* **1**, 195–200 (1965).
6. J. F. CLAUSER and M. A. HORNE, *Phys. Rev.* **D10**, 526–535 (1974).
7. H. P. STAPP, in: *Microphysical Reality and Quantum Formalism*, (A. VAN DER MERWE, *et al.*, eds.) (Kluwer, Dordrecht, 1988), pp. 367–378.
8. O. COSTA DE BEAUREGARD, *Nuovo Cim.* **B42**, 41–64 (1977); **B51**, 267–279 (1979); *Phys. Rev. Lett.* **50**, 867–869 (1983).
9. H. P. STAPP, *Nuovo Cim.* **B29**, 270–276 (1975); *Am. J. Phys.* **53**, 306–317 (1985).
10. W. C. DAVIDON, Quantum physics of single systems, *Nuovo Cim.* **B36**, 34–40 (1976).
11. J. RAYSKI, *Evolution of Physical Ideas Towards Unification* (Universitatis Iagellonicae Folia Physica, Krakow (1995).
12. C. W. RIETDIJK, *Found. Phys.* **11**, 783–790 (1981); *Nuovo Cim.* **B97**, 111–117 (1987).
13. J. G. CRAMER, *Phys. Rev.* **D22**, 362–376 (1980).
14. R. I. SUTHERLAND, *Int. J. Theor. Phys.* **22**, 377–384 (1983).
15. V. BUONOMANO, *Nuovo Cim.* **B57**, 146–170 (1980).
16. V. BUONOMANO, in: *Microphysical Reality and Quantum Formalism*, (A. VAN DER MERWE, *et al.*, eds.), Kluwer, Dordrecht (1988), pp. 67–76; *Found. Phys. Lett.* **2**, 565–568 (1989).
17. P. A. M. DIRAC, *Proc. Roy. Soc.* **A180**, 1–20 (1942).
18. R. P. FEYNMAN, *Int. J. Theor. Phys.* **21**, 467–488 (1982).
19. W. MÜCKENHEIM, *Lett. Nuovo Cim.* **35**, 300–304 (1982); *Phys. Rep.* **133**, 339–381 (1986).
20. J. MADDOX, *Nature* **320**, 481 (1986).
21. I. D. IVANOVIC, *Lett. Nuovo Cim.* **22**, 14–16 (1978).
22. J. SIX, *Phys. Lett.* **A150**, 243–245 (1990).
23. D. HOME, V. L. LEPORE, and F. SELLERI, *Phys. Lett.* **A158**, 357–360 (1991).
24. S. CASER, *Phys. Lett.* **A102**, 152–158 (1984).
25. A. GARUCCIO and F. SELLERI, *Phys. Lett.* **A103**, 99–103 (1984); F. SELLERI, *Phys. Lett.* **A108**, 197–202 (1985).
26. W. PERRIE, A. J. DUNCAN, H. J. BEYER, and H. KLEINPOPPEN, *Phys. Rev. Lett.* **54**, 1790–1793 (1985).
27. R. RISCO DELGADO, *Tesis Doctoral*, Universidad de Sevilla (1997).
28. A. ASPECT, P. GRANGIER, and G. ROGER, *Phys. Rev. Lett.* **47**, 460–463 (1981).
29. C. H. THOMPSON, *Found. Phys. Lett.* **9**, 357–382 (1996).
30. B. C. GILBERT and S. SULCS, *Found Phys.* **26**, 1401–1439 (1996).
31. L. DE LA PENA and A. M. CETTO, *The Quantum Dice: An Introduction to Stochastic Electrodynamics*, Kluwer, Dordrecht (1996).
32. M. FERRERO and E. SANTOS, *Found. Phys.* **27**, 765–800 (1997).
33. D. HUME, *Enquiries Concerning Human Understanding* (Clarendon, Oxford, 1963), quoted in Ref. 32.

34. A. EINSTEIN, *The Born–Einstein Letters* (Macmillan, London, 1971), pp. 170–171.

35. A. EINSTEIN, in: *Electrons et Photons—Rapports et Discussions du Cinquième Conseil de Physique tenu à Bruxelles du 24 au 29 Octobre 1927*, Gauthiers-Villars, Paris, (1928), pp. 254–256.

36. Quoted by M. JAMMER, *The Philosophy of Quantum Mechanics* (Wiley, New York, 1974), p. 116.

37. J. S. BELL, letter to E. SANTOS, quoted in Ref. 32.

38. F. J. BELINFANTE, *A Survey of Hidden-Variable Theories*, Pergamon Press, Oxford (1973).

39. K. POPPER, in: *Open Questions in Quantum Physics* (G. TAROZZI, *et al.*, eds.), Reidel, Dordrecht (1985), pp. 3–32.

40. D. Bedford and F. Selleri, *Lett. Nuovo Cim.* **42**, 325–328 (1985).

Index